D1287807

THE BIRDS OF DEVON

THE BIRDS OF DEVON

by

ROBERT F. MOORE

DAVID & CHARLES : NEWTON ABBOT

7153 4720 9

TO VERONICA

Printed in Great Britain by
W J Holman Limited Strand Dawlish Devon
for David & Charles (Publishers) Limited
Newton Abbot Devon

Contents

List of Plates

Preface

The last book on the avifauna of Devon was W. S. M. D'Urban and the Rev M. A. Mathew's comprehensive and authoritative work, *The Birds of Devon*, published in 1892, of which a second edition, with a supplement, was issued in 1895. Apart from being now out of date, it has long been out of print and is difficult to obtain. The *Victoria County History of the County of Devon*, published in 1906, contains a section written by D'Urban on the birds of the county, which admirably summarised his former book but included little or nothing fresh. Only the first volume, dealing mainly with the natural history of Devon, was published, and the work was never completed.

D'Urban, who resided at Newport House, at the head of the Exe Estuary, lived to the great age of ninety-seven and, until about eight years before his death in 1934, continued to keep records of the birds of the county. His interleaved and closely annotated copy of this section of the *Victoria County History*, which came into my possession through the kindness of W. Walmesley White of Budleigh Salterton, helps to bridge the gap in the records between the publication of D'Urban and Mathew's book and the issue of the first Annual Report of the Devon Bird-Watching & Preservation Society, for the year 1928-9.

The present book endeavours to summarise the main changes that have occurred in the bird life of Devon during the twentieth century and to give an account of the status of every species recorded up to the end of 1967.

ACKNOWLEDGMENTS

Using the work of D'Urban and Mathew as the foundation, my book is based principally on the forty *Annual Reports of the Devon Bird-Watching Society*, from 1928 to 1967, and the seventeen *Annual Reports of the Lundy Field Society*, covering the twenty years from 1947 to 1966; the Report for 1967 was unfortunately not available at the time of writing.

This book therefore represents the labour of a great number of people, many of whom I am happy to count as friends, who have found pleasure and relaxation in studying the birds, and almost certainly other forms of wildlife, of this most beautiful county. Many of them have readily given of their time and knowledge, in answering questions, providing information and helping in a number of other ways, and I am most grateful for their valuable assistance and advice. In particular I wish to thank O. D. Hunt, F. R. Smith and H. P. Sitters for reading and criticising all or parts of the draft; E. H. Ware and R. J. Hosking for going to great trouble to provide illustrations; Michael Blackmore, W. O. Copland and B. Grimes for help with the illustrations; R. A. W. Reynolds for the loan of books; P. J. Dare, M. R. Edmonds, P. F. Goodfellow and Tony Soper for help and advice, and the following for very kindly providing information : R. G. Adams, D. K. Ballance, H. H. Davis, C. H. Fry, G. H. Gush, C. Guy, L. I. Hamilton, R. H. Higgins, H. G. Hurrell, F. C. H. Kendall, J. D. Magee, C. G. Manning, J. L. F. Parsloe, R. G. Roper, B. L. Sage and W. H. Tucker. Not least, I thank my daughter, Helen, for patiently typing the MS, and my wife, Veronica, for encouragement and forbearance.

In thanking the Officers of the Devon Bird Watching and Preservation Society and those of the Lundy Field Society for allowing me to make such extensive use of their records, I should also like to pay tribute to the part played by these Societies, and more recently by the Devon Trust for Nature Conservation, in helping to bring about a more enlightened attitude in Devon towards the conservation of wildlife.

It is essential, however, that there should be no lessening of effort, but that everyone who has the love of the Devon countryside at heart should support the nature conservation movement in order to safeguard this precious heritage which can otherwise so easily be whittled away.

RFM
Plymouth
June 1969

The County of Devon

Boundaries

Devon, with an area of 2,605 square miles, is the third largest county in England. Occupying the central and widest part of the Devonian Peninsula, it is bounded on the west by Cornwall, on the east by Somerset and a corner of Dorset, and on the north and south by the Bristol and English Channels. The River Tamar, for most of its length, forms a natural boundary between Devon and Cornwall, while in the north-east and east, the high ground of Exmoor and the Blackdown Hills are a natural boundary with Somerset.

The greatest distances from north to south and from east to west are almost identical, being 73 miles from Foreland Point to Prawle Point and 72 from the Dorset border near Hawkchurch to the Cornish border near North Petherwin. The mainly cliff-bound coastline extends for a total length of approximately 160 miles; it is indented by the estuaries of no less than thirteen rivers and includes many fine headlands.

The centre of Devon is dominated by the granite mass of Dartmoor, rising to just over 2,000 ft and overlooking mile after mile of distant pasture lands and hedgerows, rolling hills and heavily-wooded river valleys.

The many and diverse habitats contained within these boundaries include high moorland, commons, fast-flowing rivers; deep, wooded valleys, extensive conifer plantations, rough grazing and agricultural land; inlets and tidal estuaries with their adjacent marshes and sand dunes, and a varied foreshore; all of which provide for a correspondingly rich and varied avifauna.

The necessarily brief and incomplete sketch of the main topographical features of the county which follows is intended principally to give some idea of the diversity of the environment and its influence on the bird life.

Coastline

The north coast, fringing the Bristol Channel, extends for almost 60 miles, stretching westwards from Glenthorne on the edge of Exmoor to Marsland Mouth on the Cornish border. It embraces much of the finest and most unspoilt coastal scenery in Devon, including the sheer cliffs of Foreland Point, the thickly-wooded slopes of Woody Bay and, westwards, the rugged coastline to Highveer Point and Combe Martin. This stretch includes the deep-cut valley of Heddon's Mouth and the twin peaks of Holdstone Barrows and the Great Hangman which rise to over 1,000 ft. The main sea bird colonies of Razorbills, Guillemots, Fulmars, Cormorants and Herring Gulls are situated on this part of the coast, dotted along the cliffs between Lynton and Combe Martin, together with multitudes of Jackdaws, scattered pairs of Ravens, Great Black-backed Gulls and occasional Lesser Black-backs.

Continuing west, then southwards, from Combe Martin to Saunton, the line of cliffs is broken by Woolacombe Sand, but includes the headlands of Bull, Morte, and Baggy Points. Herring Gulls breed plentifully along most of this stretch, Stock Doves are fairly frequent, and Fulmars have recently colonised a number of sites and continue to spread. Baggy Point supports breeding colonies of gulls, Cormorants and Shags. Beyond Baggy and Downend the combined Rivers Taw and Torridge flow between Braunton and Northam Burrows into the wide sweep of Bideford Bay.

After Westward Ho! the land gradually rises and the cliffs stretch almost unbroken to the storm-ravaged bare precipices of Hartland Point, and thence southwards to Marsland Mouth on the border with Cornwall. With little foreshore and scant vegetation, these flat-topped cliffs are not quite so attractive to birds but, nevertheless, provide nesting sites for Jackdaws and Herring Gulls, occasional pairs of Oystercatchers, Ravens, and Kestrels and, on the clifftops and grassy slopes, Rock Pipits, Stonechats, Linnets and Whitethroats are amongst the passerine birds that nest in the gorse and bramble patches.

Until the disappearance of the rabbit from the clifftops, Buzzards nested commonly along the north coast but have ceased to do so since the 1950s. Peregrines, too, were once a familiar sight at a number of long established eyries on both the north and south coasts, but have now become practically extinct.

In the south almost exactly 100 miles of Devon coast lie between Dorset and Cornwall. Rather more devious than the north coast, it runs mainly south-west from the Dorset border to Prawle Point, the most southerly tip of Devon, then north-west to Plymouth Sound. The greater extent of foreshore, the more shallow sea and the numerous sheltered inlets, bays, and estuaries, are generally more attractive to birds, particularly wildfowl, waders, and terns, than are the steep shores of the Bristol Channel.

Between Lyme Regis and Seaton is the Axmouth–Lyme Regis Undercliffs National Nature Reserve, an undisturbed area of 800 acres, principally of ashwood, which has grown naturally on the landslip. It is interesting ornithologically for its mainly woodland bird community, including one or two pairs of Nightingales which probably breed. The 120 species recorded on the Reserve, however, cover a wide variety of birds.

West of the landslip the coast is broken by the Axe estuary, beyond which rise the chalk cliffs of Beer Head, possibly one of the last breeding sites in the county of Rock Doves. From near Sidmouth the characteristic soft, red sandstone cliffs stretch westwards past Ladram Bay, with its cliff-breeding House Martins and eroded stacks, to the shingle ridge blocking the Otter estuary, and thence to the wide Exe estuary and Dawlish Warren, with its multitudes of ducks and wading birds.

Except for Hope's Nose, much of the coast from Dawlish Warren to Berry Head, including the Teign estuary, has been developed and is of interest chiefly for its wintering grebes, divers, and sea ducks. Hope's Nose, however, has a thriving colony of Kittiwakes and wintering Purple Sandpipers, while Berry Head and nearby Scabbacombe Head support sea bird colonies which include Fulmars, Kittiwakes, Guillemots, Razorbills, Cormorants, and Herring Gulls.

Beyond Scabbacombe is the deep inlet of the beautiful River Dart and the fine coastal scenery stretching south to Start Point, including the wide sweep of Start Bay and the shallow freshwater lagoon of Slapton Ley, an important feeding ground for migrant warblers and wintering ducks.

From Start Point westwards to Bolt Tail the cliffs are wild and rugged. Much of this stretch, with its gull colonies, Ravens, Shags, and Kestrels, and its bracken and bramble-covered slopes where Stonechats and Linnets nest, belongs to the National Trust and is completely unspoilt. The coast between Prawle Point and the towering cliffs of Bolt Head is indented by Salcombe Harbour and the Kingsbridge estuary. Prawle Point, the southernmost tip of Devon, is a vantage point for the observation of spring and autumn migrants, both passerine and sea birds, and has produced many interesting records over the years.

The last stretch, from Bolt Tail to Plymouth, is mostly rugged and cliff-bound, except for the low-lying and sandy parts between Hope Cove and Bigbury, which are gradually being developed. It is broken by the sandy estuaries of the Avon and Erme and the deeper inlet of the Yealm, with its steep, wooded shores. Wembury Point, almost at the entrance to Plymouth Sound, is remarkable for the tangled heaps of decaying seaweed, cast up by the tide, which provide a rich feeding ground for various wading birds, particularly Turnstones, many of which are present throughout the year.

Rivers and Estuaries

The Devon rivers provide many miles of different habitats, which support a variety of birds. The combined length of the thirteen main rivers alone, exclusive of their many tributaries, amounts to almost 400 miles, of which about 60 miles are tidal. The principal habitats are the high moorland bogs where the rivers rise, then open moorland, followed by steep, wooded cleaves through which the rivers flow down to agricultural land. In their lower reaches they wind through water meadows and later, becoming tidal, and often fringed with reed beds, they broaden out into shallow, muddy, or sandy estuaries, flanked with grazing marshes, sometimes with sand dunes at the mouth. Some of the estuaries, however, have stony shores and steep, wooded hills, between which the rivers flow through deep channels to the sea.

Of the thirteen main rivers, the Teign, Dart, Avon, Erme, Yealm, Plym, and Tavy rise on Dartmoor and flow southwards to the coast. The Taw, which rises within a stone's throw of the Dart, flows northwards and emerges through the combined Taw and Torridge estuary into Bideford Bay, on the north coast. The Torridge, which rises close to the Tamar in the marshy uplands of north-west Devon, follows a circuitous course before being joined by the Okement from Dartmoor, and entering the steep, wooded valley above Torrington before finally meeting the Taw at Appledore. The Exe, which drains much of Exmoor before crossing from Somerset into Devon, follows a southerly course through its thickly-wooded valley, then winds through water meadows between Bickleigh and Exeter. At Topsham it flows through dense reed beds before broadening out into its shallow, muddy estuary, a haven and rich feeding ground for many wildfowl and wading birds.

The Tamar similarly flows right across the county from north to south, dividing Cornwall from Devon and following a course mainly through agricultural country until it enters its twisting and deep wooded valley between Greystone Bridge and Calstock. From there it is tidal and flows through a sheltered estuary where it is joined by the Tavy before meeting the deep water of the Hamoaze at Plymouth.

The East and West Dart Rivers, which drain a great expanse of Dartmoor, join at Dartmeet to flow through miles of deep-cut valley. Continuing its tortuous course off the moor, through densely-wooded Holne Chase and Hembury Woods, the Dart descends to lowland Devon at Buckfast, from where it flows through agricultural country to Totnes, where it is tidal.

Amongst the birds harboured by these and the many other fast-flowing streams and rivers are Heron, Mallard, Dipper, Grey Wagtail, Sand Martin, and Kingfisher.

Foremost amongst the Devon estuaries is that of the Exe, a fairly large estuary, almost 6 miles long by 1 mile wide, through which the main channel winds between extensive mudflats, which gradually give place to sandbanks towards the lower end. With Exminster and Powderham Marshes on the west side, the River Clyst and its marshes on the north-east, and the sand dunes of Dawlish Warren at its mouth, the whole area is the most favoured locality for grebes, wildfowl, waders, and terns in the south-west. Now a National Wildfowl Refuge, it is perhaps most important for its wintering Brent Geese and about 5,000 Wigeon, but it also supports a vast number of birds of many species.

The combined estuary of the Taw and Torridge, together with Braunton Marsh, Horsey Island, and Northam and Braunton Burrows, constitutes a large and varied area, and is the principal resort of wading birds and ducks on the north coast. Although lacking the numbers that frequent the Exe, it is visited by a considerable variety of species. Braunton Burrows, the southern portion of which is now a National Nature Reserve, is notable for its extensive sand-dune system and flora, and formerly held the only breeding colony of Black-headed Gulls in the county, as well as several other interesting species. Many rare birds have been recorded in this area over the years, but for someone who knew it well before the construction of the airfield, the power station, and the road across the Burrows, the Taw estuary has lost much of its former charm and remoteness.

Ornithologically, the lower Tamar estuary is more interesting on the Cornish side, with St John's Lake and the Lynher River, but from Warleigh Point to above Weir Quay it holds a fair number of duck and waders and, in particular, is the regular wintering ground of about fifty Avocets. Up to 2,000 Golden Plover visit this part of the estuary which, with the Tavy, is also frequented by wintering Black-tailed Godwits and numbers of Wigeon.

The Dart estuary, although scenically beautiful, with its miles of steep, wooded cleaves and deep-water entrance between imposing cliffs, is not as attractive to waterfowl as the shallow estuaries of the Exe and Taw. For a number of years, however, it was regularly used by wintering Spoonbills, and the heronry at Sharpham, which existed until 1965, had been occupied since at least the early 1880s.

The Kingsbridge estuary, which also has a deep, narrow entrance, but is unique among the Devon estuaries in having no river, fans out between Salcombe and Kingsbridge into seven or eight shallow creeks. It holds a wintering population of up to about 2,000 Wigeon and numbers of Shelduck, Mallard, Teal, and waders.

Only passing mention can be made of the three small estuaries of the Avon, Erme, and Yealm, which, although lying within 8 miles of each other and superficially alike, are totally dissimilar in detail yet almost equally attractive. Purple Heron, Night Heron, and Little Egret have all been observed in the sheltered creeks of the Erme and Yealm in recent years and both the latter have heronries on their thickly-wooded banks. The Plym estuary, although ruined by industry, still manages to support a number of wading birds, with sometimes up to 300 Black-tailed Godwit, and fair-sized flocks of Dunlin, Redshanks, Golden Plover and Curlew.

The Teign estuary, despite its area of mudflats and wide foreshore, does not hold any great number of wildfowl, but has had wintering

Avocets and Spoonbills and frequently Red-breasted Mergansers. Of the two small estuaries in east Devon, the Otter and the Axe, the latter is the more important. Detailed studies during the past decade have shown that it is visited by a wide range of species during the course of a year.

Lakes and Reservoirs

Although endowed with so many fine rivers and estuaries, Devon is remarkably deficient of natural freshwater lakes and ponds and, indeed, the only one of any consequence is the large freshwater lagoon at Slapton, which is of considerable ornithological importance. The constant and increasing demand for domestic water supplies, however, has necessitated the provision of a number of reservoirs, several of which have been constructed at the headwaters of certain of the Dartmoor rivers, notably at Burrator, Fernworthy, Venford, and Avon Dam. In general these reservoirs are not particularly attractive to wildfowl because of their depth, steep sides and lack of aquatic vegetation, but the surrounding areas usually provide habitats and sanctuary for a variety of other species.

Slapton Ley, a natural freshwater lagoon over 2 miles long and varying in width from about 200 to 600 yd, is situated on the south coast between Dartmouth and Start Point. Lying at sea level behind a narrow shingle ridge, it faces eastwards into Start Bay. Its 248 acres of shallow standing water are divided into two halves by Slapton Bridge, which separates the Higher and Lower Leys. The two halves, however, are totally different in character, the northern, Higher Ley, being almost entirely covered by a dense growth of reeds, patches of willow carr (which form numerous islands), and smaller stands of other water plants. In marked contrast, the Lower Ley is predominantly open water, with reeds fringing most of the margins and dense reed beds in the two bays on the landward side. Also at the sheltered southern end are some large patches of water lily, a favourite haunt of Black Terns during the autumn.

Now a Nature Reserve, owned and directed by the Herbert Whitley Trust, with a Field Studies Centre and a Bird Observatory, Slapton Ley has attracted ornithologists since at least the end of the eighteenth century, and it was here and on the nearby Kingsbridge estuary that Colonel George Montagu did much of the research for

B

his *Ornithological Dictionary*. As a resort of wildfowl, the Ley supports a winter population principally of Mallard, Pochard, Tufted Duck, and Coot, together with smaller numbers and occasional visits of most other species of duck. The breeding waterfowl comprise Mallard, Mute Swan, Moorhen, and Coot, but Tufted Duck, Garganey, and Crested Grebe have nested sporadically, and the Ley also supports many nesting Reed Warblers and smaller numbers of Sedge Warblers and Reed Buntings, while Stonechats, Linnets and Whitethroats nest in the gorse and brambles along the seaward shore.

The area is also important as a resting and feeding ground for migrating birds of many species, particularly warblers and swallows, great numbers of which fatten on the teeming insect life before their departure in the autumn. Among the many rare birds that have been recorded there during recent years may be mentioned Scops Owl, Squacco Heron, Marsh Harrier, Crane, Great Reed Warbler, and Spotted Crake.

Burrator reservoir, a fairly large sheet of water, is situated in the Meavy valley, at approximately 700 ft, between Sheeps Tor and Yennadon Down, on south-west Dartmoor. Surrounded by a combination of conifer plantations and boulder-strewn hillsides, with scrub and some mixed woodland, this reservoir harbours a variety of species, but because of the lack of water plants, and also due to disturbance, does not support many wildfowl other than Mallard, Teal, Pochard, and Tufted Duck, with a few Goldeneyes and occasional Goosanders. A few waders including Green and Common Sandpipers and Greenshank occur in the autumn and a variety of other species has been recorded over the years. The surrounding country, with its old stone walls, derelict farmhouses, and some old timber, is attractive to Redstarts, of which a number of pairs breed. Woodlarks were formerly quite common but have become scarce during recent years.

Fernworthy, which lies at an altitude of 1,100 ft in a secluded part of the moor south-west of Chagford, impounds the waters of the South Teign. It is partly surrounded by rough moorland, but mainly by extensive plantations of conifers and a small amount of scrub. Here again the lack of cover around the margins and the absence of aquatic vegetation render the water unsuitable for many wildfowl, but a few pairs of Mallard and occasional Teal breed, and Pochard, Tufted Duck, and Goosanders, all in small numbers, appear in winter. The afforestation of the surrounding moorland, however, has attracted many woodland birds, including one or two, such as Lesser Redpoll and Siskin, which did not formerly breed in Devon. It also provides much needed sanctuary for predators, several species of which breed in the area.

Venford reservoir, on Holne Moor, south-east of Hexworthy, is extremely barren and exposed. Completely lacking in cover and waterside vegetation, it suffers, too, from considerable disturbance from motorists and is practically devoid of waterfowl. Similarly, the Avon reservoir, although in a fairly remote part of the moor and comparatively undisturbed, holds few birds other than Herring Gulls and such typical moorland species as Meadow Pipits, Carrion Crows and, in summer, Whinchats and Wheatears. The Common Sandpiper, which has almost disappeared as a breeding bird from Dartmoor during the past twenty years, nested in this locality until the early 1960s. Hennock reservoir, 3-4 miles north of Bovey Tracey, comprises three lakes, having a combined surface area of about 140 acres. Situated amongst conifer plantations at around the 800 ft contour, they also have steep sides with little vegetation, and support the usual Mallard, of which a few pairs breed, and wintering Tufted Duck, Pochard, Teal, and Coot.

Tamar Lake, a small reservoir of about fifty acres, lies across the Devon and Cornwall border, a few miles below the source of the Tamar and not far from Bradworthy. Situated amongst pasture fields and rough grazing land at about 400 ft, it is fringed on three sides by marshy ground; its shallow edges support a quantity of waterside plants and there is thick undergrowth in places on the banks. Unlike the Dartmoor reservoirs, Tamar Lake holds a number and variety of waterfowl including breeding Mallard, Coot, Moorhen, occasional Teal, and Little Grebes, while Heron and Great Crested Grebe have been known to nest. Many kinds of ducks, waders, geese, terns, and gulls occur there on migration or during spells of cold weather, and rarities have included Great Snipe, Greater Yellowlegs, Glossy Ibis, and Red-necked Grebe.

Wistlandpound reservoir, 12 miles north-east of Barnstaple, is situated in rough grassland on the edge of Exmoor. Much of the surrounding ground has been planted with conifers which, as they mature, will alter the habitat. The interesting list of species recorded there since 1955 includes Great Northern Diver, Long-tailed Duck,

Quail, Grey Phalarope, and Whooper Swan. Willow Tits, not often seen in that part of the county, have also been observed, and Whinchats are included amongst the breeding birds.

A number of other small reservoirs dotted about the county are unimportant so far as wildfowl are concerned, but they provide sanctuary for other birds and produce interesting records from time to time. Of the few small lakes and ponds, mostly on private estates, the small lake at Arlington Court is kept as a sanctuary by the National Trust and holds a heronry. Shobrooke Park, near Crediton, is of interest for the flock of eleven Canada Geese which were introduced in 1949 and have since increased and spread to many parts of the county. Other waters include Kitley Pond, Bicton, Creedy and Stover Ponds, Merton claypits, Blagdon Lake, and Beesands Ley, each different in some way from the others, and all having something of interest.

Moorlands

With the whole of Dartmoor, part of Exmoor, and a number of small commons within its bounds, Devon is well provided with moorland, despite the loss during the present century of many acres of grass moor and heathland to agriculture, forestry, and the production of china clay. Of the almost 400 square miles comprising Dartmoor National Park, much land, particularly on the fringes, is enclosed farmland, chiefly hay and pasture fields, and there are many acres of woodland. But over and above this there are at least 80,000 acres of open moorland, varying between bracken and gorse-clad common, grass moor, heather moor, bogland and stony hillsides.

The characteristic birds of open moorland are few, principally Meadow Pipits, Skylarks, and Carrion Crows, with a scattering of Buzzards and Ravens. Red Grouse are thinly distributed over the high moors where, in the central boglands, a few pairs of Golden Plover and Dunlin breed. Curlew and Snipe nest in the valley bogs, together with Reed Buntings and Mallard; Lapwings are thinly distributed over drier ground; Wheatears occur on the boulder-strewn hillsides and Ring Ouzels breed in some of the wilder valleys. Whinchats are fairly widespread but Stonechats are much more restricted. Many of the breeding birds leave the moor in the autumn and are replaced by foraging flocks of Starlings, Fieldfares, Redwings, mixed finches, Lapwings, Golden Plover, and the occasional wintering Hen Harrier.

Two species recently lost or almost lost as breeding birds on Dartmoor are the Black Grouse, through persecution, and the Common Sandpiper, evidently through disturbance at its restricted breeding sites.

Although a great deal of the south and west of Exmoor lying within the Devon border has been reclaimed and cultivated as sheep pasture, many acres of heather moorland still remain—between Dulverton and North Molton, astride the long ridge of Anstey and Molland Commons, and, in the north-west, Brendon Common.

Colaton Raleigh, Woodbury, Bicton, Lympstone, and East Budleigh Commons, which rise to a maximum height of only 560 ft and lie to the east of the Exe Estuary, are totally different in character from Dartmoor and Exmoor. A combination of heather and grass moors, with scattered pine trees and scrub, this small area attracts several interesting species, including Crossbills and Redpolls and, formerly, Nightingales. Red-backed Shrikes, once tolerably common, have become practically extinct. Although these heaths suffer excessive disturbance from quarrying, military training, and motorists, they nevertheless manage to retain some of their charm.

Great Haldon, formerly moorland, is now almost completely afforested, and Chudleigh Knighton Heath, once a most interesting area, in addition to being the Devon stronghold of the Nightingale, has been utterly ruined by development of one kind and another.

Woodlands

Of the 102,000 acres of woodland in 1968, representing just over 6 per cent of the county's area, 54,000 were hardwood and 48,000 are conifer plantations. The hardwoods are predominantly the hillside oakwoods of the river valleys, which are such a feature of the Devon landscape, but which unfortunately continue to be felled and replaced by conifers at an alarming rate. The main deciduous woods cover the steep valleys of the rivers Dart, Tavy, Tamar, Teign, Taw, Torridge, and Exe; but they include many others throughout the county, such as the valleys of the Bray, Heddon, and Lyn on the fringes of Exmoor; those of the Plym, Bovey, Avon, and Lyd; the thickly wooded banks of the Yealm and Erme estuaries; and also

Ashclyst Forest and Yarner Wood. The typical birds of these woods are Buzzards, Wood Pigeons, Jays, Tawny Owls, woodpeckers, tits, Nuthatches, Tree Creepers and, in summer, warblers and Redstarts, while a few of the Dartmoor oakwoods are now being colonised by Pied Flycatchers, which have been encouraged to breed by the provision of nest boxes. At Yarner Wood, a National Nature Reserve on the south-eastern edge of Dartmoor comprising 361 acres of mixed but mostly oak woodland, the Pied Flycatcher has become well established since 1955.

Of the conifer plantations, some 22,000 acres are owned or managed by the Forestry Commission and comprise in the main Halwill, Hartland, Lydford, Bellever, Fernworthy, Plym, Haldon, Eggesford, Molton, and Honiton forests. Some of the largest areas, notably Halwill and Hartland forests, have been planted on the reclaimed moorland and boggy uplands of west Devon; others, including Bellever and Fernworthy, are on Dartmoor; while Plym, Lydford, Eggesford, and Molton forests are in river valleys; and Haldon is largely reclaimed dry, heather moorland. The avifauna of moorland areas has definitely been enriched by afforestation as is shown by the addition of the Sparrow Hawk, Merlin, Montagu's Harrier, Chiffchaff, Siskin, and Redpoll as breeding species on Dartmoor, while several other species have increased in numbers, though one or two probably only temporarily. The Grasshopper Warbler and Whinchat, for instance, are attracted only to the young plantations. Wherever situated, however, the state forests are valuable reserves for many forms of wildlife.

Agriculture

Devon is basically an agricultural county; by far the greatest part of the land is devoted to farming, principally the grazing of cattle and sheep, with arable farming taking second place. The endless miles of thick hedgerows, interspersed with elm, oak, beech, ash, holly, and other trees, which divide and shelter the lush pastures, provide cover and nesting sites for the common birds of the county. In addition, the farm ponds, strips of marginal land, rushy fields, small copses, orchards, and all the other odd corners with thistles, brambles, thickets of blackthorn, and patches of gorse and scrub, harbour a great number of birds of many species.

Mechanised methods of agriculture, however, and the need to till every available yard of land, are slowly but surely changing the pattern and altering the look of the countryside. Ponds and ditches are filled in, hedges are bulldozed out of existence, marginal land is reclaimed and ploughed, wet fields are drained, and hedgerow trees are felled so as not to impede mechanical trimmers. The old familiar orchards are being grubbed up and stretches of moorland are cleared and fenced.

The decrease of the Woodlark may be associated with the loss of marginal land; the Corncrake is already lost as a breeding bird, almost certainly due to changes in agricultural practice. The Lapwing is becoming scarce as a breeding species, as is the Partridge, and the Moorhen is beginning to vanish from some of the farms, though its decrease may as yet be scarcely noticeable.

Perhaps more serious still, the widespread use of toxic chemicals as pesticides has had a disastrous effect on predatory birds at the far end of the food chain. The Peregrine, for example, has been almost exterminated; the Sparrow Hawk suffered a sharp decline, and the Barn Owl, a totally beneficial bird to agriculture, is gradually becoming more scarce.

With these gradual changes in land use and the pressing demand for land from so many quarters, the county fortunately has an active conservation body, the Devon Trust for Nature Conservation, which has already secured a number of valuable reserves, and works in close co-operation with national and other bodies.

Lundy

Lundy, a colossal block of granite, 3 miles long by roughly half a mile wide, lies at the mouth of the Bristol Channel, 11 miles north of Hartland Point, the nearest mainland, and 20 miles due west of Morte Point. Now owned by the National Trust, the island is a sanctuary for all forms of wildlife, its 1,100 acres comprising pasture fields, rough grazing, and moorland, with a marshy area and pond. In places the bare, jagged cliffs of the west coast, facing the storms of the Atlantic, rise a sheer 400 ft from the sea, their line broken here and there by grass-covered slopes and deep gullies. The rather less precipitous east coast, with its terraces and thickets of rhododendron, willow copses, and brambles, is much less exposed and, with

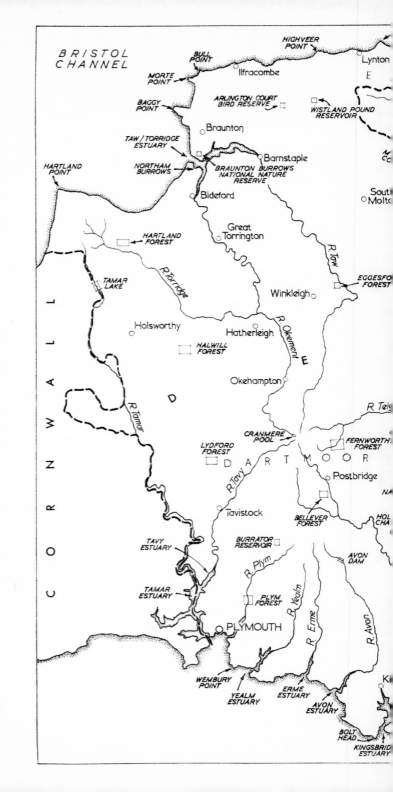

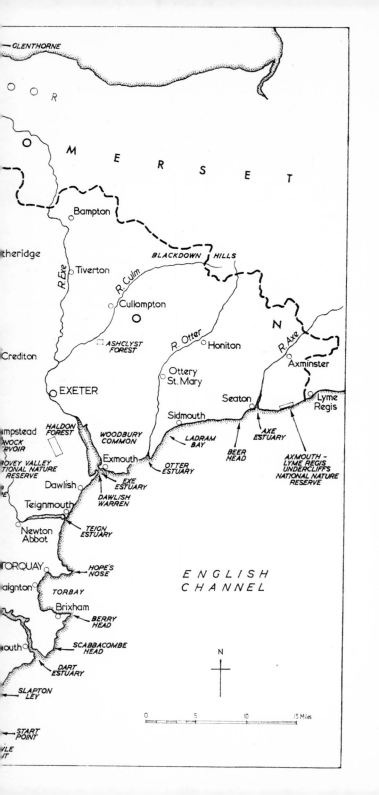

the sycamore wood in Millcombe, affords shelter to migrants that pass along the island in spring and autumn.

Ornithologically, Lundy is important mainly for its breeding colonies of seabirds, which include Kittiwakes, Great and Lesser Black-backed Gulls, Herring Gulls, Razorbills, Guillemots, Shags, Fulmars, and a remnant of the formerly abundant Puffins. During the past fifteen years it has lost its breeding Peregrines, and more recently Buzzards have ceased to nest, while Ravens have increased. Other breeding birds are Lapwing, Oystercatcher, Kestrel, Carrion Crow, Wood Pigeon, Rock and Meadow Pipits, Skylark, Chaffinch, Linnet, and others. Species which breed irregularly include Manx Shearwater, Cormorant, Curlew, Swallow, Wheatear, and Stonechat.

Following the establishment of the Bird Observatory in 1947, continuous observation by the Warden and visiting members of the Lundy Field Society has revealed that a number of passage migrants, previously unrecorded, occur regularly on the island. These include Lapland and Ortolan Buntings, Woodchat Shrike, and, irregularly, Red-breasted Flycatcher, and Barred, Melodious and Icterine Warblers.

The list of vagrants, incorporating several species reliably recorded for the first time in the British Isles, includes a number of North American species, of which may be mentioned: Myrtle Warbler, Yellowthroat, Rufous-sided Towhee, Baltimore Oriole, American Robin, Black-billed Cuckoo, Semipalmated and Least Sandpipers. South European and Asiatic species include Spanish Sparrow, Sardinian Warbler, and Bimaculated Lark, to name but a few.

The County List and Changes in the Avifauna

In the *Victoria County History* D'Urban listed 294 numbered species, but these include three doubtful occurrences (King Eider, Spotted Eagle and Black-headed Warbler) as well as six subspecies, thus making a revised total of 285 full species. The present list contains 336 species, representing an increase of fifty-one. The main additions during the present century are the Willow Tit, a resident which was not then separated from the Marsh Tit; the Collared Dove, which has colonised Britain since 1952; the introduced Canada Goose and Red Grouse, the latter introduced or re-introduced to Dartmoor and Exmoor since 1900; the Roseate Tern, which now occurs regularly on passage migration; and the Marsh Warbler, which has nested on at least two occasions.

Of the remaining forty-five species added, the Aquatic, Icterine, and Barred Warblers, Red-breasted Flycatcher, and Ortolan and Lapland Buntings have been found to occur as scarce but fairly regular passage migrants; the Little Ringed Plover and Mediterranean Gull have both been recorded on a number of occasions, and eleven New World species, the Killdeer, Greater and Lesser Yellowlegs, Semipalmated Sandpiper, Bonaparte's Gull, Black-billed Cuckoo, American Robin, Myrtle Warbler, Yellowthroat, Baltimore Oriole, and Rufous-sided Towhee, have been recorded as vagrants from America, as were the Yellow-billed Cuckoo and Two-barred Crossbill recorded last century, but not accepted at the time.

The remaining twenty-four species, all vagrants, are mostly larks, thrushes, warblers, finches, and buntings, chiefly from southern Europe and Asia, and including Bimaculated Lark, Bonelli's Warbler, and Spanish Sparrow.

Thirteen species included by D'Urban have not been reliably recorded during this century: Wilson's Petrel, Cattle Egret, American Bittern, Black Stork, Ruddy Shelduck, Great Bustard, Long-billed Dowitcher, Ivory Gull, Great Black-headed Gull, Whiskered Tern, Pallas's Sandgrouse, Alpine Accentor, and Parrot Crossbill, plus the Yellow-billed Cuckoo and Two-barred Crossbill. Thus, 321 species are known to have occurred in Devon during the present century.

Of the 146 species recorded as having bred in the county since 1900, 112 breed regularly, including Montagu's Harrier, Peregrine, Hobby, Merlin, Ringed Plover, Golden Plover, Dunlin, and Red-backed Shrike, but these eight breed in small or very small numbers and two, the Red-backed Shrike and possibly the Peregrine, may soon be lost as breeding species; the Hobby, however, has become quite well established during the past two decades. The remaining thirty-four species comprise the Gannet, Rock Dove, and Chough, which have since been definitely lost as breeding birds, although the Gannet still occurs commonly; the Black Grouse, Corncrake, Common Sand-piper, and Corn Bunting which are almost lost; and the following twenty-seven species which have bred sporadically : Great Crested Grebe, Storm Petrel, Manx Shearwater, Teal, Garganey, Shoveler, Tufted Duck, Red Kite, Red-legged Partridge, Quail, Water Rail, Woodcock, Redshank, Black-headed Gull, Common Tern, Long-eared Owl, Short-eared Owl, Wryneck, Golden Oriole, Marsh Warbler, Dartford Warbler, Black Redstart, Hawfinch, Siskin, Twite, Crossbill, and Tree Sparrow. Of these, the Redshank and Black-headed Gull became established as breeding birds for about 25 and 45 years respectively during this century, but have since been lost, the Redshank perhaps only temporarily following the extermination of the breeding stock in the severe winter of 1963. The Siskin, which has recently begun to breed, may soon become established.

Definite gains as regular breeding species are : Fulmar, Canada Goose, Red Grouse, Collared Dove, Little Owl, Lesser Whitethroat, Pied Flycatcher, and Lesser Redpoll. Species which previously bred but have increased during this century include : Shelduck, Buzzard, Curlew, Great Black-backed Gull, Herring Gull, Kittiwake, Wood Pigeon, Stock Dove, Raven, Carrion Crow, Magpie, Jay, Redstart, Reed Warbler, Starling, Goldfinch, Bullfinch, and House Sparrow. On the other hand, the Sparrow Hawk, Partridge, Lapwing, Razorbill, Guillemot, Puffin, Cuckoo, Barn Owl, Nightjar, Woodlark, Nightingale, and Ring Ouzel have all decreased as breeding birds.

Some significant increases amongst non-breeding species include the Black-tailed Godwit, formerly a vagrant, which is now a common winter visitor and passage migrant; the Spotted Redshank, Greenshank, and Ruff as winter visitors, formerly only passage migrants; the Avocet as a well established winter visitor; and the Gadwall and

Eider as winter visitors.

With the cessation of collecting, several species of wading bird occur much more commonly on the estuaries nowadays than at the close of the nineteenth century. These include: Redshank, Curlew, Sanderling, Bar-tailed Godwit, Knot, Turnstone, and Oystercatcher.

The Bluethroat, Aquatic Warbler, Melodious Warbler, Icterine Warbler, Red-breasted Flycatcher, Woodchat Shrike, Ortolan Bunting, and Lapland Bunting, which are nowadays recorded as annual but scarce passage migrants, were probably formerly overlooked.

Devon has always been well off for birds of prey, and there can be few other counties which include amongst their breeding species seven diurnal birds of prey, namely: Buzzard, Sparrow Hawk, Montagu's Harrier, Hobby, Peregrine, Merlin, and Kestrel; yet all these nested in the county in 1967, while for good measure the Rough-legged Buzzard, Goshawk, Red Kite, Hen Harrier, Osprey, and Red-footed Falcon were also recorded in that year.

By comparison with Devon, Palmer and Ballance, in *The Birds of Somerset*, list 294 species, of which about 140 have bred, and at least 116 do so regularly. There is no up-to-date list of the birds of Cornwall published; if there were, it would certainly include a large number of vagrants, particularly from the Isles of Scilly.

Systematic List

This list includes the 336 species recorded in Devon since 1800. Doubtful records have, as far as possible, been entirely excluded, as have those of the Red-headed Bunting, which is no longer admitted to the British list, and the Mandarin Duck, which has not yet been accepted. I have also omitted the Eagle Owl, Dusky Thrush, Desert Wheatear, and Rustic Bunting which, although included in the *Devon Reports*, appear to have been accepted on insufficient evidence, except in the case of the Eagle Owl, which had apparently escaped from captivity. The records of the Dusky Thrush and Desert Wheatear were not subsequently included in *The Handbook*.

The scientific names are those used in the Revised Edition (1966) of Peterson, Mountfort and Hollom's *A Field Guide to the Birds of Britain and Europe*, and the sequence of species is that of the British Ornithologists' Union's *Check-List of the Birds of Great Britain and Ireland*, published in 1952. Lack of space has prevented me from including as many ringing recoveries as I would have liked, and from comparing the status of each species in Devon with its status in the adjoining counties of Cornwall and Somerset. Place names are generally in accordance with the Ordnance Survey.

The following abbreviations have been used:

Devon Report	The Annual Reports of the Devon Bird-Watching Society
Lundy Report	The Annual Reports of the Lundy Field Society
BTO	British Trust for Ornithology
SBO	Slapton Bird Observatory
LBO	Lundy Bird Observatory
D & M	D'Urban & Mathew's *The Birds of Devon*
VCH	*The Victoria County History of Devon*
BSED	Loyd's *Birds of South-East Devon*
BL	Davis's *List of the Birds of Lundy*
BB	*British Birds Journal*
The Handbook	Witherby's *The Handbook of British Birds*

In order to make the list as complete and up-to-date as possible, I have added a few 1968 records which are of particular interest.

BLACK-THROATED DIVER   *Gavia arctica*

Winter visitor and passage migrant

The least common of the three divers that visit Devon, the Black-throated Diver occurs as a scarce and irregular winter visitor and passage migrant. Fifty-six examples have been recorded in 23 of the 40 years from 1928 to 1967. Except for two single birds seen at Lundy in November 1951 and April 1964, one on the Taw estuary in March 1936, and a storm-driven bird at Boyton in September 1935, all have occurred along the south coast. Of these, seventeen were observed off Dawlish Warren, thirteen in Start Bay, eight on the River Yealm, and the remainder at various other places.

Eleven occurred during January, nine in February, sixteen in April and up to five in each of the months of March, May and September to December. The April and May records are indicative of a small passage movement along the south coast, observed in some years, and include the occurrence of four off Dawlish Warren on 21 April 1953, of which two were in breeding plumage. Other than this, all have occurred singly or in twos, and many of the spring birds were in breeding plumage.

Judged by D & M's account of this species, there appears to have been no change in its status since the nineteenth century.

GREAT NORTHERN DIVER   *Gavia immer*

Winter visitor

A regular winter visitor, the Great Northern Diver occurs off both coasts but is much more frequent in the south where singles and scattered parties of up to five or six, often more, are reported from all localities from Seaton to Plymouth. For the most part they keep to the sea, occurring everywhere along the coast, but quite often inside the estuaries and inlets, and sometimes on freshwater.

Although in some years, due to the presence of non-breeding individuals, the species is recorded in every month, the majority occur from November until March, with early arrivals in September and October, and late birds, not infrequently in summer plumage, during May.

On the north coast, although singles are recorded fairly regularly from the Taw and elsewhere, it is normally much less common, but large numbers have exceptionally been encountered, as for example the loose flock of about 120 which A. J. Vickery saw off Baggy Point on 27 January 1963, of which all those near enough to be identified were of this species. Similarly, about forty were seen in this locality in December 1964. Reports of singles from inland waters include the occasional occurrence at Tamar Lake, Wistlandpound, Meldon, and Burrator.

Records of more than usual on the south coast include : twelve off Exmouth on 23 December 1950, twenty-five off Dawlish Warren on 30 October 1955, twelve at Branscombe on 3 January 1956, about thirty-eight moving eastwards off Sidmouth on 7 January 1959, sixteen off Exmouth on 24 December 1961, eighteen at Dawlish Warren on 11 December 1965, and nine at Beesands in January 1965 and again in January 1966.

Apart from a few old records, singles or up to three have been reported from Lundy in about ten of the years from 1947 to 1966, mostly in spring and autumn.

RED-THROATED DIVER   *Gavia stellata*
Winter visitor

A regular winter visitor in varying numbers, the Red-throated Diver, although occurring annually in very small numbers on the north coast and the Taw and Torridge estuary, is much more plentiful in the bays and estuaries of the south. Whilst occasional birds are recorded as early as September, they more often arrive during December, but stay on well into March and April, and sometimes May, by which date some have assumed breeding plumage.

Like the other divers, this species does not normally congregate into flocks but is generally dispersed along the entire coast, singly or in loose gatherings of up to six or seven, occasionally many more. Although infrequent on inland waters, singles have been noted several times on Burrator and Tamar Lake, and very occasionally at Shobrooke, Merton, Squabmoor, and elsewhere, while it occurs regularly on freshwater on the Exeter Canal and sometimes travels several miles up the rivers, as at Bishops Tawton and the River Tavy.

The principal records listed in the *Devon Reports* are : sixteen off the Otter Estuary on 26 December 1948, ten at Slapton in January 1953, thirty-eight off Dawlish Warren on 3 January and nineteen

Page 33: *Curlew at nest. A widely distributed breeding bird, nesting principally on moorlands and rough hill pastures*

Page 34:
A corner of Braunton
Burrows, a locality
favoured in winter
by Short-eared Owls.
Until the 1960s it
held the only Devon
breeding colony of
Black-headed Gulls

on the Exe above Topsham on 14 March 1956, at least fifteen in Chiselbury Bay in December 1963, thirteen at Beesands on 18 January 1965, and the unusual number of eighty-three recorded by S. C. Madge between Slapton and Beesands on 23 January 1966.

This species occurs irregularly off Lundy, where about fifteen, mostly singles, have been observed in ten of the years since 1939, usually in spring and autumn.

The status in Devon appears to be unchanged since last century.

GREAT CRESTED GREBE *Podiceps cristatus*
Winter visitor, breeds sporadically

D & M, who knew of no breeding record of this species, regarded it as a frequent winter visitor. During the present century it is recorded as having bred on at least ten occasions and suspected on several others, between 1930 and 1945. It is stated in *BB* 26:72 that a pair reared three young at Tamar Lake in 1930 and that Slapton Ley was colonised in 1930 and one pair nested there in 1931. *BB* 30:153 records that at least one pair bred at Slapton in 1935, and a pair bred at Blagdon Lake, Ashwater, in 1933 and 1934. The *Devon Reports* confirm the latter nesting at Ashwater, and also record that several young were seen at Slapton in 1934, that two pairs nested there in 1940, and a pair was displaying in April and a juvenile was seen there in September 1945. Adults were observed at Slapton in May 1936 and June 1937.

As a winter visitor it occurs regularly in small parties of up to four or five in the bays and estuaries of the south coast, chiefly from November to March, but it has been recorded in all months. In the north ones and twos occur irregularly on the Taw estuary and Tamar Lake, and it has been reported occasionally from several other coastal localities and inland waters. There is also an isolated occurrence at Burrator in February 1959, but it has not been observed at Lundy.

Records of more than the usual numbers include twenty-five in Torbay in December 1947; groups of fifteen or sixteen off Dawlish Warren in January 1949, March 1950, January 1952, January 1957, and January 1959; eleven at Paignton in January 1962; over forty at Plymouth during the severe cold of January/March 1963; up to

C

twenty-six at Paignton in January/February 1964; and up to seventeen there in January 1966.

RED-NECKED GREBE   *Podiceps grisegena*

Irregular winter visitor

This, the rarest of the five grebes that occur in Devon, has never been more than a scarce and irregular winter visitor, its status being much the same now as during the last century. It has been reported in 22 of the past 40 years, the records involving about thirty-eight birds, two-thirds of which occurred during January, February and March, and the remainder from October to December, with one each in August and September.

Almost always occurring singly, but with three on the Exe estuary on 5 February 1967, this species has been reported most frequently at Dawlish Warren on the Exe estuary. Twenty-three of the records refer to this locality, two or three each to Slapton Ley, Tamar Lake, the Taw estuary, and Plymouth, and one each to Torbay, Start Point, the Yealm and Axe estuaries, and Lundy. The bird observed at Lundy from 23 to 31 August 1957 constitutes the first and only record for the island, and is an unusually early date.

Possibly due to more observers rather than an actual increase in the number of birds, the Red-necked Grebe has been reported slightly more regularly during recent years, and annually since 1961.

SLAVONIAN GREBE   *Podiceps auritus*

Winter visitor

The Slavonian Grebe occurs as a regular winter visitor along the south coast, mainly on the sea, but often inside the estuaries during rough weather, and occasionally on inland waters. It is nowadays rare on the north coast and the Taw, where D & M formerly regarded it as the commonest grebe.

Occurring between November and April, sometimes in October, it is most frequent off Dawlish Warren, but also appears annually in smaller numbers in Torbay and Start Bay and occasionally in Plymouth Sound and the other south coast estuaries.

The main records for the Exe are : fourteen in February 1946 and March 1952, fifteen in December 1953 and November 1961, sixteen in January 1963, twenty-two in January and thirty-two in November 1964, twenty in November 1965 and December 1966, and twenty-two in March 1967. The peaks in March may well include some passage migrants. Maximum numbers for other localities include thirteen on the Kingsbridge estuary in December 1946, and nine at Paignton in November 1963, ten in January 1965, and twelve in January 1967.

Singles were reported at Burrator reservoir in January 1947 and December 1963, at Hennock in March 1955, and ones and twos have several times occurred at Tamar Lake. The few records for the Taw and Torridge estuary relate to a pair during the winter of 1934-5 and one in February 1950. It is likewise rare at Lundy, where four were seen during April 1947, and singles have occurred in November 1951, October 1955, and March 1957.

BLACK-NECKED GREBE   *Podiceps nigricollis*

Winter visitor

The Black-necked Grebe, described by D & M as a casual winter visitor, is nowadays regular in winter in some of the sheltered bays and river mouths of the south coast, but is rare on inland waters and on the north coast.

At the mouth of the Exe, where it has been recorded annually since the early 1930s, it became quite numerous during the 1940s and 1950s, when counts of up to twenty were not unusual, and a maximum of thirty-nine was recorded by F. R. Smith on 9 March 1952. Reports of other larger than usual numbers include twenty-eight in December 1950 and February 1953, twenty-nine in December 1953, and twenty-seven in February 1954. The first arrivals usually appear in early November, sometimes September or October, but the peak occurs from December to February, while small numbers, often in summer plumage, are still present in early April, and may include passage birds.

In recent years this species has become much less frequent on the Exe but rather more plentiful in Torbay, where the annual number now varies from about six to eleven, with a maximum of possibly twenty in February 1962. Ones and twos are recorded annually at Slapton and irregularly on the Yealm, Plymouth Sound, the Kingsbridge estuary, and elsewhere. On the Taw estuary ones and twos are very occasionally reported, and inland occurrences have been recorded at Melbury, Dalditch and Holsworthy reservoirs, Tamar Lake, and Bicton.

The only fully substantiated record for Lundy refers to one, in company with a Slavonian Grebe, on 1 November 1951.

LITTLE GREBE   *Podiceps ruficollis*

Resident and winter visitor, breeds

Although a fairly common winter visitor to all the estuaries, the Little Grebe is scarce as a breeding bird because of the shortage of suitable quiet ponds and slow-flowing rivers with adequate vegetation. Breeding occurs, however, at several scattered localities, but in smaller numbers since the severe winter of 1963, which took a heavy toll of this species.

Localities at which breeding has been reported during the past twenty years include: Tamar Lake, where one or two pairs breed annually; Horsey and Wrafton Ponds; the lower reaches of the Rivers Otter, Axe, Erme, Dart, and Teign; the old canal at Halberton; the ponds at Buckland Filleigh, Creedy, Kitley, Chudleigh Knighton, and Merton; Exmouth and possibly Slapton Ley. Nesting may also occur on the Rivers Clyst and Culm, and elsewhere.

As a winter visitor, it occurs principally on the estuaries, but also in small numbers on most of the reservoirs, including Burrator, Fernworthy, and Wistlandpound, and occasionally on the sea. The greatest numbers resort to the south coast estuaries where they are relatively common from October to March, though not now in the numbers recorded prior to 1963. At Newton Ferrers, O. D. Hunt recorded at least 100 on the Yealm in January 1949, and 30-60 in many winters. Fifty were counted on the Kingsbridge estuary in February 1959, but ten to twenty is an average for most estuaries.

This species is rare on Lundy where singles were reported in October 1949, April 1963, and August 1964.

BLACK-BROWED ALBATROSS   *Diomedea melanophris*

Vagrant

Devon can claim one record of this accidental straggler from the southern oceans. A bird identified as an immature Black-browed

Albatross was observed off Morte Point on 25 April 1965 by M. E. Greenhalgh. In his description of the bird, given in the *Devon Report* for 1965, the observer states that it was seen at a distance of 200 yd. The record was accepted by the Rarities Committee and published in *BB* 59:283 with a footnote to the effect that this is the third British record and that the preceding years, 1963 and 1964, provided three records of albatrosses in Irish waters, one in each year being identified as Black-browed.

WILSON'S PETREL   *Oceanites oceanicus*
Vagrant

Although it is not recorded in *The Handbook* as having occurred in Devon, an example of this rare vagrant is stated by D & M to have been obtained at Exmouth on 13 November 1887. As D'Urban subsequently handled the bird and confirmed the identification, there seems to be no good reason for not believing it to be a perfectly satisfactory record. Two other, unsubstantiated, records are given by D & M and a further, rather unsatisfactory record, is contained in the *Ilfracombe Fauna and Flora* which states that 'one was obtained locally in 1899 . . . was probably obtained from the Taw'. This petrel breeds in the Antarctic and 'winters' in the north Atlantic during our summer, but occurs only as a vagrant in Britain.

LEACH'S PETREL   *Oceanodroma leucorrhoa*
Irregular winter visitor

The occurrences of this pelagic species on the coast of Devon, or more rarely inland, is invariably due to continuous westerly or south-westerly gales which drive these small sea-birds in from the Atlantic. D & M's account shows that Leach's Petrels were recorded in about fifteen of the years between 1823 and 1891, all the occurrences being in the months of September to January, with the majority in November. Whilst two or three of these records refer to the north of the county, most occurred in the vicinity of Plymouth. Except for the 'wreck' of 1891, the numbers involved have usually been small and frequently single birds, but D & M state that many were obtained after the severe storms of 1891.

Similarly, the recorded occurrences for the present century are few and mostly relate to individual birds that have been driven inshore by autumnal gales. On 11 December 1948, however, at least six birds were seen by R. G. Adams on the Exe estuary after a week of southerly gales; and one was picked up at Exmouth on 10 December and another at Newton Abbot on 15 December of that year.

In the catastrophic 'wreck' of Leach's Petrels which took place in the British Isles between 21 October and 8 November 1952, at least sixty-three birds were reported from various parts of the county, many of these from the coast between Plymouth and Start Point, a few individuals inland at Yelverton, Tavistock, Holsworthy, and Bridgerule, and a number on the Taw estuary and the north coast. On the Taw estuary 50-100 birds were seen on 25 October, but of these there is no record of how many were Leach's and how many Storm Petrels. Twenty-eight Leach's Petrels were reported flying over the Taw at Chivenor on 29 October. Great numbers of this species were driven up the Bristol Channel and, of the total 'wreck' involving some 7,000 birds, about a third were reported from Bridgwater Bay, in Somerset. A detailed account of this 'wreck' was given by H. Boyd in *BB* 47 : 137-63.

Although many Leach's Petrels must have passed close to Lundy during October and November 1952, none was actually recorded there. The only record for the island is the remains of a bird of this species found on 2 June 1928.

Four of the five occurrences since 1952 refer to single birds on the south coast in the months of September and October. The fifth relates to a bird seen by Dr Elliston Wright near Braunton, at intervals between December 1959 and February 1960.

STORM PETREL   *Hydrobates pelagicus*

Pelagic, summer visitor to land, has bred

Except when gale-driven on to the coast, the Storm Petrel is seldom seen from the shore, though it occurs regularly a few miles out to sea, mainly between April and October. Although recorded in all but two years since 1947, the numbers observed are usually small, but twelve were encountered during a crossing from Ilfracombe to

Lundy on 25 July 1954, instead of the usual two or three. In June 1953, however, R. M. Lockley saw several hundred to the north of Lundy.

North-westerly gales in the summer often bring a few close inshore at Ilfracombe, where six were observed in June 1947.

On the south coast small parties of up to six have frequently been observed during the summer, a few miles out to sea from Exmouth. In 1959 twelve were seen on 17 and sixteen on 29 October three miles off Start Point, and it is regular off the Eddystone Light, where, amongst other records, fifteen were noted on 19 August 1952. On 29 October 1964 seven struck the lighthouse at Start Point and, during a gale on 14 October 1967, ten were observed off Hope's Nose. Gale-driven birds have been picked up at Okehampton, Shirwell, and Lustleigh.

The 'wreck' of Leach's Petrels in October 1952 included 50-100 Storm Petrels which entered the Taw estuary.

Although D & M stated that this species occasionally breeds on Lundy, it has not been proved to nest and is seldom seen on the island, but dead birds were found there in June 1956 and July 1961 and 1962.

Breeding on the Thatcher Rock, Torbay, was recorded in about 1874. The *Devon Report* records that a pair bred there in 1950, when an adult, incubating a fresh egg, was handled.

MANX SHEARWATER   *Puffinus puffinus*

Resident and partial migrant, breeds

The Manx Shearwater, an essentially maritime species, occurs commonly off the coasts from mid-March or early April until October and sometimes November.

The often large movements eastwards and westwards along the north coast, frequently observed from Ilfracombe during May to July, are evidently fishing flights, presumably of birds from the colonies off Pembrokeshire; similarly, a movement north of some 15,000 birds was seen off Hartland Point by O. D. Hunt on 10 July 1963. Smaller numbers occur regularly off the south coast during the summer, but a considerable passage westwards is observed off Start and Prawle Points mainly in September/October. Birds have also been observed coming ashore at Start Point during the summer, but there is no evidence of breeding.

Although several thousand have occasionally been observed off Lundy in June and July, and varying numbers occur regularly from March to September/October, the numbers actually breeding are very small. Davis states 'breeds in several small colonies, but many of the birds that come ashore are non-breeders . . .' *The Handbook* records that eggs were taken on Lundy in 1903 and 1920-22. Proof of breeding was also obtained in 1942 and young were seen in 1934-5 —BB 38 : 122-9. The *Lundy Reports* indicate that although breeding has been attempted in probably most of the years 1947-66, it is usually unsuccessful, possibly because of rats. Successful breeding occurred in 1959, and in August 1966 the remains of several young were found near a Great Black-backed Gull's nest.

The Balearic race *P.p. mauretanicus* is recorded annually off the south coast, with many more than usual in 1955 when counts of 256, 120, and 140, mostly of this race, were observed off Prawle Point on 13/14 September.

GREAT SHEARWATER *Puffinus gravis*

Irregular visitor

D & M refer to large numbers of this species sometimes occurring off the south coast, usually well out to sea, and mention that many occurred in November 1874. Details are given of about ten examples obtained off the south coast during the nineteenth century. D'Urban quotes that Eagle Clarke found it common in the vicinity of the Eddystone Light in September and October 1901. G. M. Spooner recorded in the *Devon Report* for 1945 that considerable numbers were reported in the Channel off the Eddystone Light during October and November, and a few in December, and that two which were caught and brought to him proved to be this species.

In 1965 J. R. Brock observed a single example of this or Cory's Shearwater about a mile off Prawle Point on 5 September, and one was reported between the mainland and Lundy on 16 July 1966. R. Perry in *Lundy, Isle of Puffins* mentioned seeing two off the island on 14 April 1939, and the *Lundy Report* for 1950 records that one or two were seen close to the island on 20 and 21 April, and two were encountered between Lundy and Bideford on 6 September of the same year. A south Atlantic species, the Great Shearwater 'winters' in the north Atlantic during our summer.

SOOTY SHEARWATER *Puffinus griseus*

Vagrant

Of this rare shearwater which breeds in the southern hemisphere, D & M were able to quote only two undated records of singles obtained at Plymouth. Eagle Clarke observed three or four singles off the Eddystone Light in September and October 1901. D'Urban in the *VCH* refers to one seen by E. A. S. Elliot off Bigbury in May 1900, an unusual date as this species normally occurs in the autumn.

The *Devon Reports* contain six or seven records, the first of which relates to two birds reported off Ilfracombe on 28 May 1942, but lacks evidence of identification and is unusually early. In 1965 three were observed off Slapton on 17 and one on 24 September, and two off Beesands on 26 November, while in 1967 one was seen off Slapton on 13 September and singles, possibly the same individual, off Start Point and Stoke Point on 7 October.

This species has not been recorded from Lundy.

FULMAR *Fulmarus glacialis*

Pelagic, summer visitor to land, breeds

The Fulmar in D & M's time was no more than a vagrant to Devon, having been reliably recorded only once in the north and five times on the south coast. Until it began to prospect Lundy in 1938 and eventually bred for the first time in south-west England in 1944, there were only about three other recorded occurrences in Devon, one at Thurlestone in November 1898 and singles off Lundy in June 1921 and 1935.

The breeding population on Lundy increased steadily during the 1950s; in 1958 sixteen young were successfully reared, and the population was stated to be about thirty-five pairs in 1959, but only twenty-seven in 1965. The birds are present at Lundy from about February until September, with irregular occurrences during the remaining months.

The first mention of the Fulmar in the *Devon Reports* was in 1944 when the Lundy breeding was quoted from *BB* 38:97-8. Breeding first occurred on the mainland in 1949, at Berry Head, where prospecting birds had been regularly observed since 1946. It was subsequently proved at Scabbacombe in 1950, and on the north coast the first egg was found at Martinhoe in 1956. Prospecting birds were meanwhile observed at many other sites along practically the entire north and south coasts during the 1950s, and in increasing numbers during the 1960s.

In 1967 established breeding colonies were known at Baggy Point, Bennett's Mouth, Ilfracombe, Berrynarbor, Combe Martin, Martinhoe, Woody Bay, and east of Lynton, on the north coast, and on the south coast at Berry Head and Scabbacombe, with a maximum of about fifty birds at Berry Head in May. It continues to increase and spread.

Fulmars are present at their breeding cliffs from about January or February until September, but for a shorter period at the sites still being prospected.

GANNET   *Sula bassana*

Regular non-breeding visitor, formerly bred

The Gannet is known to have bred on Lundy from the thirteenth century. *The Handbook* records that there were sixteen nests in 1887, about seventy in 1889, and thirty pairs present in 1893. Adults were present until about 1907 but the last known eggs were taken in 1903. A single bird attempted to build in 1922 and possibly in 1927. The decline of the Lundy gannetry during the nineteenth century coincides with the colonisation of Grassholm, about 45 miles distant.

This species nowadays occurs regularly off the mainland and Lundy and is recorded in all months of the year, the largest numbers occurring off the south coast during the autumn and winter. Smaller flocks, mainly on fishing flights, occur off the north coast, often after westerly gales, and the species tends to become less frequent eastwards, up the Bristol Channel; but 180 were recorded off Combe Martin on 2 September 1966. It is regular and frequent off Lundy, where the maximum numbers of up to about 100, usually much less, occur during August and September, occasionally October.

On the south coast, the greatest numbers are usually seen after westerly gales, but fishing flights occur at all times, and the pattern of visits tends to vary from one year to another. The irregularity of the peak numbers is illustrated by the following records: Sidmouth 200 during December 1933 and January 1934; Start Point 250 on 29 January 1936; Wembury 2,000 on 4 February 1942; Exmouth 200-300 on 7 October 1945; Teignmouth 300 on 2 November 1952; Start Bay almost 500 on 16 January 1953; Dawlish Warren 300 on 13 December 1956; Prawle Point 300 in September 1963; Stoke Point 800 and Dawlish Warren 400 in gale movements, both recorded on 14 October 1967.

In the south few or none are recorded in some years between February and April or May, in others, between June and July, and the Gannet, although occurring annually, is much less common in some years, as in 1959 when the largest number noted was only thirty-one.

CORMORANT   *Phalacrocorax carbo*

Resident, breeds.

A numerous resident, the Cormorant occurs commonly in the estuaries and adjacent shallow seas; it is regular on freshwater and travels far up the rivers, as well as visiting the Dartmoor reservoirs. The numbers are augmented in winter, when birds ringed on the Irish and Welsh coasts have been recovered in Devon.

On the north coast breeding occurs at Baggy Point and at several sites between Combe Martin and Lynton, while in the south it has been recorded at Ladram Bay, Brandy Head, between Dawlish and Teignmouth, Scabbacombe, Dartmouth, Start Point, east and west of the Erme, and probably elsewhere, but nesting may not be regular at all sites because of the toll taken by fishermen. In 1967 thirty-one nests were counted at Brandy Head, fifty pairs were reported at a colony east of Plymouth, and five pairs at Scabbacombe.

In winter up to 100 birds are regularly counted on the Exe estuary where a maximum of 177 was recorded in November 1965; the greatest number recorded in recent years was a flock of 280 off Teignmouth on 17 October 1947, and 132 were counted on the stacks off Hope's Nose in August 1965. Numbers of up to about fifty are frequently reported from other localities including Slapton. Overland migration across Dartmoor has been observed in both spring and autumn.

Breeding occurred regularly on Lundy until 1955 and intermittently until 1959, but not since. The breeding population has decreased gradually from sixteen pairs in 1932 to eight in 1948 and one in 1959. Peaks of up to about thirty still occur on spring and autumn passage, but it was always scarce in winter.

SHAG   *Phalacrocorax aristotelis*

Resident, breeds

The Shag, a resident but more maritime species than the Cormorant, appears to have decreased slightly since D & M's time, although the numbers have fluctuated over the years. Small colonies are scattered along both the north and south coasts, where the birds breed on precipitous cliffs. In the north, it now nests regularly on Baggy Point and at a number of sites between Combe Martin and Lynton. Along the south coast it breeds in small numbers at Ladram Bay and Brandy Head, but in larger numbers from Torbay westwards, at Hope's Nose and Scabbacombe Head, and from Start Point to Bolt Tail. There is also a colony on the Mewstone at Wembury.

Recorded numbers of nests in recent years include fifteen at Wembury in 1960, twenty-five at Bolt Tail and eleven at Start Point in 1961, and eighteen at Start Point in 1967. Larger than usual gatherings include thirty at Baggy Point in August 1955, fifty-seven at Wembury in August 1957, and over 100 at Start Point in June 1961. The Shag occurs infrequently inside the estuaries and rarely on freshwater, but an assembly of 134 was recorded on the Exe estuary during rough weather in November 1945, and a single bird was reported on Melbury reservoir in September 1965.

On Lundy, where it greatly outnumbers the Cormorant, the Shag occurs mainly from March to September and in small numbers throughout the remaining winter months. The breeding population was recorded by Loyd in 1922 as about twelve pairs, by Perry as 110 breeding pairs in 1939, and in the *Lundy Reports* as fluctuating from thirty-seven pairs in 1950 to 132 in 1956. The number was stated to be ninety pairs in 1963, but only fifty-three in 1965.

GREY HERON *Ardea cinerea*
Resident and winter visitor, breeds

Resident and mainly sedentary, the Heron is well distributed through-out the county, occurring on all the rivers, estuaries, inland waters, and on quite small streams, but is most frequent on the estuaries. The population is reduced by hard winters but usually recovers within a few years and, overall, remains fairly constant, although it does not yet appear to have recovered from the exceptional winter of 1963. Until then it was fairly frequent on Dartmoor where from 1948 or earlier and until 1962 up to five pairs bred at Archerton, at 1,300 ft. Since 1900 breeding has been recorded at a number of other localities on Dartmoor, including Bagga Tor, Dunnabridge, Soussons, and Hexworthy, while at present three pairs breed at Neadon Cleave on the River Bovey. Foraging birds occur at bogs far out on the moor, and have been recorded at Cranmere Pool.

Breeding is mainly concentrated in about ten heronries, listed in the table overleaf, most of which are on estuaries and have been in existence since at least the last century. In some instances, however, the exact sites have changed slightly, often due to the felling of trees. The long-established heronry at Powderham moved in about 1910 to Eastdon House, returning later to Powderham. The old site at Sharp-ham, on the Dart, mentioned in 1830 by Dr E. Moore, was deserted in the mid-1960s. The total of 135 occupied nests counted in 1967 is probably incomplete and omits one or two small, scattered heronries.

Among the many places at which breeding has occurred since the 1930s are: Annery House, Slapton Ley, Clifford Bridge, Tamar Lake, Countess Wear, Dittisham, Walreddon, Maristow, Bickleigh, Hols-worthy, Shobrooke, Saltram, Landcross, Bradford, Stoke Canon, and Topsham, while in 1963 a pair bred on Drake's Island in Plymouth Sound. That some immigration occurs in the winter is shown by the recovery at Blackawton in March 1948 of one ringed as a nestling in Sweden in June 1947. Records of coastal movements include a flock of eleven flying south-west at a great height over Dawlish Warren on 19 September 1960.

Although it does not breed on Lundy and was formerly considered rare, singles and parties of up to three have been recorded annually since 1947, in all months but mainly from July to September, with maxima of nine in July 1940 and six on 6 September 1953.

DEVON HERONRIES

Locality	River	No. of occupied nests			Remarks
		1928	1954	1967	
Killerton	Culm	40	32	16	Listed by D & M
Powderham	Exe	18	31	23	do
Halwell	Kingsbridge	9	14	7	do
Stedcombe	Axe	9	19	10	do
Arlington	Taw	8	41	14	do
Puslinch	Yealm	4	5	9	do
Stiddicombe	Avon	11	7	5	Originally Stadbury Wood
Tracey	Otter	-	10	12	Established c 1919
Hele Bridge	Torridge	-	3	16	do c 1954
Warleigh	Tavy	20	?	10	Not recorded 1954
Sharpham	Dart	14	15	-	Listed by D & M Last nest, 1 in 1965
Orcherton	Erme	8	5	?	Listed by D & M Not counted in 1967
Various small sites	.	34	43	13	Listed by D & M
Total nests		175	225	135	
No of sites		16	22	15	

PURPLE HERON *Ardea purpurea*

Vagrant

Of the six or seven records of this species listed by D & M relating to four or five birds obtained and two seen, the latest and in their opinion the best authenticated refers to a bird shot on the Tamar on 30 October 1851. More than a century passed before the Purple Heron was again recorded in the county, when in 1960 an immature example was present on the marshes at Saltram, Plymouth, from 7 to 15 October. It was first seen by L. I. Hamilton whose full account appeared in the *Devon Report* for 1960, and subsequently by a number of observers. Since then there have been two spring occurrences of single birds: an adult at Puslinch on the River Yealm, from 25 April until 10 May 1965, first seen by myself and R. M. Moore, and an almost adult bird on the nearby River Erme from 27 April to 10 May of the following year, seen by S. C. Madge, O. D. Hunt, and L. H. Hurrell. These two records are fully documented in the *Devon Reports* for 1965 and 1966 and were accepted by the Rarities Committee, *BB* 59:283 and 60:312. The latter journal lists seven British records for 1966 of this summer visitor to the Netherlands and southern Europe.

A further occurrence, accepted by the Rarities Committee, refers to one at Slapton Ley 10-24 August 1968, which was seen by M. R. Edmonds and others.

LITTLE EGRET *Egretta garzetta*

Vagrant

The number of recent occurrences in Devon of this vagrant from the Mediterranean leads one to suppose that some of the earlier records, which were rejected as probable escapes from captivity, did in fact refer to genuine wild birds. D & M list four or five records, only one of which was accepted by the editors of *The Handbook*, namely a bird killed on the River Exe near Topsham on 3 June 1870, given in error in the *Birds of Devon* as May 1878.

During the present century the Little Egret is known to have occurred in Devon on at least twelve occasions, the majority of them during the past two decades; in fact there is only one reported occurrence in the first half of the century. As the records show, this distinctive and very easily identified species nowadays occurs almost annually, usually during April, May, and June, always singly, and with two exceptions, always in the south of the county. The accepted records are as follows:

1930   7 August, Axe estuary (W. Walmesley White), BB 24:131
1951   18-20 April, Erme estuary (O. D. & D. B. Hunt)
1951   17 June, Axe estuary (Lady B. Drewe)
1953   16-24 May, Otter estuary (G. H. Emerson)
1957   22-26 April, Lundy (LBO)
1958   14-21 May, Kingsbridge estuary (R. H. Stephen &
       D. R. Edgcombe)
1959   24 November to 17 December, Yealm estuary
       (L. I. Hamilton)
1960   20 May, River Clyst at Topsham (R. F. & R. M. Moore)
1961   12 June, Budleigh Salterton (R. H. Baillie)
1961   16 June, Axe estuary (F. R. Smith)
1961   1-10 May, Braunton Pill, Taw estuary (A. J. Vickery &
       D. Wilson)
1962   May, Salcombe (E. Payne)
1965   3-10 April, River Erme (R. Burridge & J. Harris)
1965   23 April, Slapton, probably the same bird (SBO)
1966   11-17 June, Axe estuary (R. Cottrill)
1967   14 June, Axe estuary (R. Cottrill)

The 1957 occurrence is the first and only for Lundy; the bird was found dead on 26 April, probably from starvation. This occurrence is detailed in the *Lundy Report* for 1957; the others are fully documented in the *Devon Reports* for the years concerned.

SQUACCO HERON   *Ardeola ralloides*

Vagrant

According to D & M five examples of this south European Heron were obtained in the county during the nineteenth century. Of these, two were said to have been killed on the Tamar and one at Kingsbridge, but they are not fully documented. The only adequately labelled specimens are a bird shot at Blatchford, near Ivybridge, in June or July 1840 and an adult male killed on Braunton Marsh on 10 June 1878.

Page 51: *Buzzard at nest. This species breeds throughout the county and,*
after the Kestrel, is the commonest bird of prey

Page 52:
The West
Dart river,
above
Hexworthy.
A typical
Dartmoor
river, flowing
through
moorland,
rough grazing
land, and both
deciduous
and coniferous
woods

Of the three or four records of this small heron for the present century, the first relates to a bird shot on the Exe estuary on 17 May 1920 by a boatman from Powderham, and reported in *BB* 14:234 by T. P. Backhouse. The second and third, which may refer to one individual, are sight-records of a bird seen at Slapton from 1 to 9 June by M. R. Edmonds and later seen on the River Exe near Bampton on 17 July 1958 by Dr Ingram. A very full account of this occurrence is given in the *Devon Report* for 1958, and the record was accepted by the Rarities Committee and listed in *BB* 53:159. The most recent record, and the only autumn occurrence as far as is known, refers to a bird in first winter plumage which frequented Slapton Ley from 19 to 30 September 1964. It was first seen by F. R. Smith and subsequently by many observers, including myself, and was filmed by H. G. Hurrell and K. Watkins. Described in detail in the 1964 *Devon Report*, it was also listed in *BB* 58:355 after acceptance by the Rarities Committee.

CATTLE EGRET   *Bubulcus ibis*

Vagrant

Of the very few authentic records of the Cattle Egret, or Buff-backed Heron, in the British Isles, the first refers to a bird shot at South Allington in October 1805, which passed into the hands of Colonel Montagu who recorded the occurrence in his *Supplement*. Writing of it as the Little White Heron, he relates that 'we had the honour of announcing this species for the first time as British, in the *Transactions of the Linnean Society*, a female having been shot near Kingsbridge, the latter end of October 1805. . . .' This specimen, now over 150 years old, may still be seen, together with many other birds from Montagu's collection, at the Natural History Museum, South Kensington. There is no subsequent record for Devon of this almost cosmopolitan species.

NIGHT HERON   *Nycticorax nycticorax*

Vagrant

The obituaries of about fifteen Night Herons shot in Devon between 1844 and 1876 are listed by D & M, who regarded this species as a

D

casual visitor in spring and autumn. Most of these were obtained on the estuaries of the south, including no less than four pairs of adults which were shot on the Erme during May and June 1849, but one, an adult male, comes from north Devon, having been shot on the Taw near Barnstaple in May 1869. The *VCH* adds two further victims: a male shot near Countess Wear in June 1897 and another killed near Kingsbridge in April 1899. The visits of this beautiful crepuscular heron during the present century have been rather more infrequent and amount to seven or eight occurrences. D'Urban's MS records that a male, shot at Newton St Cyres on 25 June 1912, was presented to the Exeter Museum. L. R. W. Loyd in *BB* 12 : 238 records that one was killed at Whitford Bridge on the Axe near Seaton on 5 November 1918. The *Devon Report* for 1928-9 states that three were seen on several occasions during June and July 1928 near Greystone Bridge on the Tamar.

On 30 October 1936, following strong westerly winds in the North Atlantic, an immature bird in a fresh condition was found floating at the mouth of the Yealm by O. D. Hunt. On examination, the bird was found to be of the American race (*Nycticorax n. hoactli*). This record, which appeared in the *Devon Report* for 1936, seems to have been either rejected or overlooked by the editors of *The Handbook*, as it is not mentioned. A Night Heron which frequented the River Avon below Loddiswell from 27 March until 4 May 1954 may possibly have been one of several that escaped from Edinburgh Zoo, as birds were reported in six different counties round about this time. The most recent records relate to single birds seen on the Otter estuary by R. H. Baillie on 21 April 1960, at Chelson Meadow, Plymouth, by R. Burridge on 4 April 1965, and at Puslinch on the Yealm by P. Harris on 29 April of the same year. These records were considered to relate to genuinely wild birds, as stated in *BB* 59 : 284. This species has not been recorded on Lundy.

LITTLE BITTERN   *Ixobrychus minutus*

Vagrant

Described by D & M as a casual visitor usually in spring and autumn, the Little Bittern is recorded by them as having occurred in the county on about twenty occasions during the last century, mostly in the south-west and generally in the spring. The *VCH* states that one of these minute bitterns was shot near Newton St Cyres during April 1901, and another at Hatherleigh early in April 1903. It is strange that this species, a summer visitor breeding as close as France and Holland, should nowadays be so rarely encountered in Devon, where it has been listed only once in the records of the county bird-watching society. This occurrence refers to a single bird which frequented the River Tavy at Denham Bridge, where it was seen by a number of observers including H. G. Hurrell and G. M. Spooner, who recorded the fact in the *Devon Report* for 1934. This bird, which was present during April 1934, was also seen by M. J. Ingram who reported it in *BB* 28 : 53.

BITTERN   *Botaurus stellaris*

Winter visitor

A scarce but nowadays an almost annual winter visitor, the reed-loving Bittern has been reported on about fifty occasions during the past 40 years, mostly during cold spells, and in the months of December to February, but with records from late September to April.

Of these occurrences, five are from Braunton Marshes, seven from the Exe estuary, mostly in the Topsham area, two from Tamar Lake, and ten from Slapton Ley. It is noteworthy that the ten Slapton records all refer to the years from 1960 onwards, including two birds seen there on 13 March 1965, and one can only suppose that its visits to the dense reed beds of the Higher Ley during the previous three or four decades, and before the area was so regularly watched, were undetected. The remaining twenty-five records are from all parts of the county, mainly coastal and estuarine areas, but including one from a typical Dartmoor bog near Sheeps Tor, another from Sticklepath, near Okehampton, and a few from north-west Devon. The only recorded occurrence of the Bittern on Lundy relates to one caught, photographed and released on 24 September 1930, and reported in *BB* 25 : 217.

D'Urban in the *VCH* summarised the status of this species as 'a winter visitor of rather irregular appearance, coming in flights at long intervals', and it is stated in the Supplement to D & M that at least a dozen were believed to have been shot at Slapton in the winter of 1891. D'Urban's MS contains records of a further twenty occurrences in various parts of the county from 1898 to 1927.

AMERICAN BITTERN   *Botaurus lentiginosus*
Vagrant

The American Bittern, a somewhat smaller bird than its European counterpart, has been recorded only twice in the county, the birds having been obtained on both occasions. The first was shot at Mothecombe on the Erme estuary on 22 December 1829, and the second was killed on a moorland near Parracombe towards the end of October 1875.

It was the Devon ornithologist, Colonel George Montagu, who first described this species, under the name of the Freckled Heron, from a specimen obtained near Piddletown in Dorset in the autumn of 1804 which came by chance into his possession a few years later.

WHITE STORK   *Ciconia ciconia*
Vagrant

The White Stork, always a rare visitor to the south-west, has been recorded twice in Devon during the present century, compared with five or more occurrences during the previous one. The old records, listed by D & M, refer to three examples obtained at Slapton Ley between the years 1820 and 1830, one shot on the Exe estuary at Topsham on 28 July 1852, and another shot near Clyst St George, probably at the mouth of the Clyst, in January 1856. White Storks seen at the last-mentioned locality, at Aveton Gifford, and on the Kingsbridge estuary, during September 1898, were believed to have escaped from captivity.

Loyd refers to one that was reported in the *Field* as having been seen on the Exe in 1913. The only record contained in the *Devon Reports* relates to a bird of this species which was seen on Dartmoor from 19 to 21 June 1949 by A. B. and G. E. May, but whether a genuine migrant or an escaped bird, cannot now be said.

BLACK STORK   *Ciconia nigra*

Vagrant

Although Bannerman, in his magnificent *Birds of the British Isles,* follows Witherby and previous writers in ascribing to Devon one of the twenty-five accepted occurrences of the Black Stork in this country, exactly where this particular bird was shot is uncertain and it may have been just outside Devon. As D'Urban points out in the *VCH*, the example shot on 5 November 1831 was killed either on the Tamar or else on the Lynher river, a tributary of the Tamar on the Cornish side. In addition to the foregoing, two sight-records, neither of which is very convincing, are mentioned by D'Urban : first, a bird supposed to have been seen on the Exe near Topsham on the rather unlikely date of 12 February 1855 in severe weather, and, second, two birds which were reported to have been observed at Salcombe Cove, Sidmouth, on 3 April 1894. D'Urban himself appears to have been somewhat sceptical of the latter record, the circumstances of which are given in the Supplement to his *The Birds of Devon.*

The Black Stork has some connection with Devon, however, because its first reported occurrence in the British Isles was made known by Colonel Montagu, who kept a specimen alive at Kingsbridge for over a year after it had been captured on Sedgemoor, Somerset, on 13 May 1814.

SPOONBILL   *Platalea leucorodia*

Winter visitor

The *VCH* states that about thirty Spoonbills were recorded during the nineteenth century, mostly immatures and mainly during the autumn and winter. D'Urban's MS and *BB* both record several occurrences of ones and twos during the first quarter of the present century, and *BB* 21 : 236 records that nine visited the Axe estuary for a few days from 28 September 1927.

The *Devon Reports* contain records for twenty-nine of the years from 1934 to 1967, during which it has occurred annually since 1942. Most of the records are from the estuaries of the Teign and Dart, but many also from the Axe, Exe, and Tamar, and less frequently from the Otter, Tavy, and the Taw and Torridge. As there is much movement of birds between the south coast estuaries, it is impossible to state how many individuals are involved, but taking the maximum number seen together each year at least 100 must have occurred since 1934. The majority were recorded during the months from September to April, but there are frequent records of non-breeding birds during May to July.

The Spoonbill was most common during the 1940s and 1950s when parties of up to four or five were frequently seen, but since 1960 only ones and twos have occurred except in 1962 when two pairs of adults were seen on the Tamar during May, and up to two on the Teign during both winter periods. The principal numbers were observed during 1951, with seven on the Teign in January, eleven in October, and fourteen on the Oar Stone in Torbay on 16 October. There were seven on the Teign in December 1952, eight in January, and six during the autumn of 1953.

This species has not been recorded on Lundy.

GLOSSY IBIS   *Plegadis falcinellus*

Vagrant

Of about fifteen occurrences of the Glossy Ibis listed by D & M, the last was on the River Dart near Totnes on 26 October 1869. This species has been recorded on some five occasions during the present century, four times during the autumn and once in the winter. These five records all refer to single birds, the first of which was reported in *BB* 3:230 as having occurred near the Taw and Torridge estuary during the last week of October 1909. The next refers to a bird seen by W. Walmesley White on the Exe estuary on 24 September 1920, and listed in the first *Devon Report*. This bird is recorded in *BB* 14:138 as having also been seen by T. P. Backhouse on 23 September and subsequently shot near Topsham on 27 September, when it was given to the Exeter Museum. In the *Devon Report* for 1942 Dr F. R. Elliston Wright mentions that one was shot on Braunton Marsh on 1 September of that year. The fourth record for this century refers to a Glossy Ibis seen at Tamar Lake on 25 September 1959, and what was almost certainly the same bird seen by three observers at Braunton Marsh two days later. This occurrence is reported in great detail in the *Devon Report* for 1959. The latest record is that of a bird which frequented the River Axe between Axminster and Axmouth from 21 November until 19 December 1964, when it was seen to leave in a south-westerly direction. This record *(Devon Report 1964)* was accepted by the Rarities Committee in *BB* 58:357, in which they add that the Glossy Ibis may be expected to occur less frequently in Britain as the number breeding in western Europe continues to decline.

What was most probably the Axe estuary individual was reported on the River Camel at Wadebridge in Cornwall from December 1964 until March 1965.

MALLARD   *Anas platyrhynchos*

Resident and winter visitor, breeds

A resident breeding species and an abundant winter visitor, the Mallard is widespread throughout the county from the estuaries and coastal marshes to the central moorlands. As a breeding bird it is present on all inland waters, pools, and rivers, and on Exmoor and Dartmoor, where scattered pairs nest along the rivers and in valley bogs. The main breeding grounds, however, are the estuarine marshes, particularly those of the Exe and the Taw and Torridge; and Slapton Ley, and the marshes along the lowland reaches of the rivers. The moorland breeding areas are deserted during the winter.

As a winter visitor, the Mallard occurs on all the estuaries and the main inland waters, with the greatest concentrations on the Exe estuary which holds an average winter population of about 700 and a maximum of 1,200 in September 1945 and December 1962; Slapton Ley with a regular number of 300 or less and a maximum of about 700 in September 1956; the Taw and Torridge with 100-300; about the same numbers on the Yealm and on the combined Plym, Tavy and Tamar; and smaller numbers on the other estuaries. Large numbers occur from time to time on the sea in Chiselbury Bay where a maximum of about 800 was noted in November 1962, and at least 500 in December 1960, while 500 were recorded at Shobrooke Park in September 1959.

From one to three Mallard occurred irregularly on Lundy between 1947 and 1957, in which year a pinioned pair was introduced. A small, mainly residential, flock since raised from this pair has confused the recording of subsequent wild birds.

TEAL *Anas crecca*

Resident and winter visitor, breeds occasionally

D & M regarded the Teal as a tolerably numerous winter visitor which had frequently nested at Slapton, had been known to breed at Beesands, and was thought to breed on some of the Dartmoor bogs.

Breeding was reported at Braunton Marshes in 1907, 1930, 1932, 1947, and 1955 when two broods were seen; a nest was found at Slapton in 1932. On Dartmoor, young were seen at Burrator in 1930 and 1932, and breeding was suspected at Tavy Cleave in 1935 and at Knattabarrow Pool in 1942; it was also suspected in the Erme valley in 1938. Two pairs with young were seen at Kitley Pond in 1948, and young were seen in the same year at Fernworthy where breeding was also suspected in 1952. A pair was thought to have bred at Tamar Lake in 1953 and probably other years; W. Walmesley White stated that it used to breed on the Otter Marshes.

The Teal occurs principally, however, as a winter visitor in fluctuating numbers to estuaries, inland waters, marshes, and flood waters throughout the county, but is infrequent on the sea. A few are recorded in August, with numbers building up during the autumn to peaks in December to February, and thereafter decreasing during March and early April, with some evidence of passage during March.

Except for flocks of up to about fifty on the reservoirs, this duck is infrequent on Dartmoor, and in the Postbridge area occurs only as an occasional winter visitor. Some of the larger numbers recorded in Devon are: Exe estuary 1,000 in winter 1950-1, 1,200 in January 1961; Taw estuary 500 in February 1957; Yealm estuary 360 in February 1956. *Wildfowl in Great Britain* gives the regular number of 390 for the Exe estuary and 155 for the Taw estuary.

Very small numbers occur almost annually on Lundy, where it has been recorded in all months except June, with a maximum of thirty on 31 January 1954.

Examples of the American form, the Green-winged Teal *(A.c. carolinensis)* have been recorded on three occasions: Kingsbridge 23 November 1879, Exe estuary 26 April 1953 *(BB* 47 : 83), and Kitley Pond March/April 1962.

GARGANEY *Anas querquedula*

Passage migrant, has bred

A regular passage migrant, the Garganey occurs in small numbers from mid-March to mid-April, less frequently on autumn passage in August, exceptionally in winter, and has bred sporadically. It is a freshwater duck, frequenting shallow pools with abundant vegetation, and occurs principally at the head of the Exe estuary, on Exminster Marshes, Slapton Ley, Braunton Marshes, Tamar Lake, and irregularly elsewhere.

D'Urban remarked in the *VCH* that it had become very rare and had not been recorded for the past 20 years. Except for occasional records, it was not until the 1940s that the Garganey was regularly observed, but since 1946 it has been recorded every year except 1954.

During March/April 1959 unprecedented numbers occurred, with a peak of over twenty on Exminster Marshes on 23 April, six each on the Plym and Tamar Lake, twenty-one on the Taw estuary, and others elsewhere. At least one pair remained and nested successfully on Exminster Marshes, where F. R. Smith saw a duck with young on 14 June. In the spring of 1960 twelve adults were recorded on the Exe, where three pairs bred successfully. Breeding was proved for the first time at Slapton in 1961 by my finding a nest beside the Higher Ley on 2 May which later hatched successfully. Breeding may have occurred at Slapton in 1962, but has not since been recorded in Devon.

Other unusual records include a bird shot at Branscombe on 6 December 1947, a drake seen at Slapton on 27 December 1949, a flock of eighteen observed at Bicton Lake on 12 March 1941, and nine on the Plym Marshes on 20 March 1964. The species has twice been recorded on Lundy where a drake occurred 24-31 March 1958 and parties of up to three during the influx of March/April 1959.

GADWALL *Anas strepera*

Winter visitor

A bird of freshwater rather than the coast, the Gadwall has increased since the time of D & M, who considered it very rare and knew of only about seven occurrences during the entire nineteenth century. Although still uncommon, it nowadays occurs as a regular winter visitor, having been recorded annually since 1944, but on only three occasions between 1928 and 1943.

Always in small numbers, it is now regular on the Exe and at Slapton Ley, and probably at Tamar Lake, where it has been seen on many occasions; there are also occasional records from most of the inland waters, including the Dartmoor reservoirs, and fairly frequent occurrences on the Taw and Torridge estuary. The visits are mainly during the months from November to February, with some from September onwards, and until early April.

Records of numbers exceeding the usual small parties of up to four or five include : ten on the Exe estuary in September 1944, six at Slapton in January 1952, six at Beesands Ley in November 1956, up to sixteen on the Exe during the winter of 1960-61, nine in the winter of 1961-2, seven in December 1963, six at Slapton in January and February 1963, and seven in December 1964.

For Lundy there are only three records which relate to ones and twos seen in July 1947 and April and May 1952.

WIGEON   *Anas penelope*

Winter visitor

The Wigeon, a rather more maritime species than the Mallard, is by far the commonest duck in Devon, occurring as an abundant winter visitor, chiefly on the estuaries but also on freshwater at Slapton and irregularly in small numbers on inland waters including the Dartmoor reservoirs. D'Urban, who considered it greatly reduced in numbers at the close of last century, said that thousands still sometimes occurred at Slapton and Thurlestone Leys.

The main resort now and since at least the 1930s is the Exe estuary where 4-6,000 have been recorded annually for the past thirty years or more, with a maximum count of 7,000 on 26 December 1961. *Wildfowl in Great Britain* gives the regular number for the Exe as 3,785 and maximum 6,000, while for the combined Tamar, Tavy, and Plym the figures are 665 and 1,535, for the Kingsbridge estuary 1,875 and 3,065, Slapton 90 and 760, and the Taw and Torridge 770 and 2,300, with regular numbers of up to about 100 on the Yealm, Erme, Avon, and Axe estuaries.

The numbers on Slapton Ley have decreased in recent years, but 3,500 were reported there in November 1956, and in February of the same year a maximum of 3,500 was recorded on the Taw estuary.

The first autumn birds arrive on the Exe during August and September, following which the flocks build up to their peak in November or December, thereafter decreasing during January and February in mild winters, but less quickly in cold weather, when many remain until the end of March and some until early April. Occasional individuals stay throughout the summer but the species has never been known to breed in Devon.

Ones and twos are recorded on Lundy most years, but not annually. Although more frequent in the autumn, it has been reported in all except summer months, with a maximum of six in February 1956, and possibly nine in January 1955.

PINTAIL *Anas acuta*

Winter visitor

The elegant Pintail, which is now a regular and tolerably common winter visitor, was stated by D & M to have become one of the rarest of the wildfowl and occurring only occasionally. Its principal resort in Devon is the Exe estuary where it occurs from October onwards until March, reaching a total of usually over 100 in the months from December to February.

Wildfowl in Great Britain gives the average for the Exe as seventy and the maximum as 185, while the main records from the *Devon Reports* during the past twenty years are 158 in February 1947, 160 in February 1953, 150 in February 1954, 125 in February 1956, and 167 in January 1964, with over 100 in a number of other years since 1947, including 105 on floodwater on Exminster Marshes in November 1960.

The Pintail, although not numerous on other waters, has been reported occasionally on most of the inland waters, including Burrator, but not the other reservoirs on Dartmoor; fairly regularly, if not annually on Slapton Ley and the Taw and Torridge estuary in numbers of up to about twenty; and irregularly from every other estuary. It also occurs on the water meadows of the Exe at Stoke Canon.

This freshwater duck has been recorded only twice on Lundy, where singles occurred in December 1932 and September 1964. It has been known to breed once or twice in both Somerset and Cornwall, but not in Devon.

SHOVELER   *Anas clypeata*

Winter visitor, has bred

D & M knew of no breeding record of the Shoveler, which they considered to be a winter visitor, not uncommon in some years. It is recorded in *BB* 2 : 53 that a pair bred at Braunton in 1904 and possibly in the two succeeding years. D'Urban's MS states that a pair bred at Thurlestone Ley in April 1904, and Dr Elliston Wright listed it as breeding at Braunton.

As a winter visitor the Shoveler occurs rather erratically and in small numbers on most of the estuaries between the months of September to April. Its principal resorts are the Exe, Taw, and Tamar estuaries, Slapton Ley, and Tamar Lake, but there are also very occasional records from most inland waters, including Burrator, Fernworthy, and Avon reservoirs.

Although the greatest numbers are recorded on the Exe and Slapton Ley, they vary there from as few as five in some winters to fifty to sixty in others. The main records for the Exe are thirty-five in December 1945, fifty-five in January 1947, and forty-three in December 1959, while at Slapton sixty occurred in January 1956 and forty-three in December 1959. Other records include twenty-six at Tamar Lake in November 1963, thirty-two on the Erme estuary in January 1955, and fifty at Camels Head Creek, Plymouth during the extreme cold of January 1963.

Davis quotes three dated records for Lundy up to 1949, since when a single occurred in March 1964.

RED-CRESTED POCHARD   *Netta rufina*

Vagrant

The Red-crested Pochard, a south European species which was formerly regarded as a rare vagrant to the British Isles, has occurred with far greater frequency in Britain in recent years, evidently as a result of the north-westwards extension of its range on the Continent during the past twenty or thirty years. *The Handbook* admits only two occurrences for Devon, the first of which relates to an adult drake shot on Braunton Burrows at the end of December 1867, while the second is presumably the example that was in the small collection of T. S. McLaughlin, the boatman at Powderham, who was stated by T. P. Backhouse in *BB* 14:162 to have shot it on the Exe estuary in November about the year 1910. The former record was the only one known to D & M.

There is no further report of this very handsome diving-duck until 1952 when a female was present on Tamar Lake for three weeks from 5 January, being reported by G. H. Martin and F. E. Carter. In the same year a drake was observed by O. D. and D. B. Hunt at Slapton Ley, on 21 December; both occurrences are included in the *Devon Report* for 1952. A duck was again present on Tamar Lake from 8 to 30 January 1953, where it was seen by F. R. and A. V. Smith. In view of the possibility of its being an escaped bird, enquiries were made of the Wildfowl Trust who considered it to be a genuine wild bird. An adult drake which occurred on Slapton Ley from 29 to 31 March 1956 was reported by M. R. Edmonds and D. R. Edgcombe. A female seen many times on the Plym estuary and also at Kitley Pond on the nearby Yealm during the years from 1959 to 1962 was considered to be an escape (*BB* 53:414). The two most recent records, both of ducks, refer to one seen with a flock of Pochard at Wistlandpound reservoir by A. J. Vickery on 4 and 10 October 1964 and one on the Plym estuary on 16 and 17 April 1966.

SCAUP *Aythya marila*

Winter visitor

A regular winter visitor in small numbers, the Scaup occurs annually on the Exe estuary and Slapton Ley, and occasionally on some of the other waters, including the Taw and Kingsbridge estuaries and Tamar Lake. Most years fewer than ten are reported for the whole county, but more occur in severe winters such as 1963 when up to nineteen frequented the Exe from early January until mid-March and six were reported at Slapton during February, with a few others elsewhere. Other larger than usual numbers include nine on the Exe during February 1953, sixteen on the Kingsbridge estuary on 11 February 1956, thirteen at Slapton on 24 December 1957, twelve each at Slapton and the Exe during October 1959, and seven on the Kingsbridge estuary in December 1962.

The visits of this duck are usually between October and March, but they have been recorded in April and May in quite a number of years. Ducks and immature birds predominate, but adult drakes are reported most years. The only two reliable records for Lundy are an oiled bird on 9 and 10 October 1955 and one at a small pond from 5 to 10 May 1965.

Although D & M described the Scaup as being 'one of the most numerous of the wildfowl . . . in the autumn . . . and keeping in large flocks some miles off the coast', there are no records of such gatherings during the present century.

TUFTED DUCK *Aythya fuligula*

Winter visitor, has bred

Although the Tufted Duck has bred sporadically during the present century, it is principally a winter visitor, though it occurs in all months. Like the Pochard, with which it frequently consorts, it occurs mainly on lakes and reservoirs and, to a lesser extent, estuaries but is rarely encountered on the sea.

The most favoured localities are Slapton Ley and Tamar Lake, with smaller numbers at Beesands Ley, and Burrator, Fernworthy and Hennock reservoirs, but it is recorded irregularly at every inland water and the lower reaches of the main rivers, and is much more widespread in hard winters.

The main numbers occur between October and March, with the highest counts in November to February. *Wildfowl in Great Britain* gives the average for Slapton as 105 and the maximum as 200, but 300 were counted there in November 1950 and again in December 1961, while for Tamar Lake the average is stated as forty and the maximum count was about 130 in the winter of 1956-7. Fairly large numbers were present in the severe winter of 1962-3 when 180 frequented the Dart, over 100 were on the Kingsbridge estuary, 150 were at Slapton, and small parties occurred on several rivers and coastal creeks. Up to ten are annually recorded on the Exe estuary where a maximum of 106 was reported during cold weather in February 1954.

It is recorded in *BB* 26:230 that breeding occurred at Slapton in 1931 and about four pairs bred in 1932. The only other record refers to presumed breeding at Tamar Lake in 1961, when a family party was observed in early July.

The Tufted Duck has been reported about five times on Lundy, where singles occurred in April 1938, October 1941, October 1943, April 1963, and March 1966.

POCHARD *Aythya ferina*

Winter visitor

A freshwater duck, the Pochard occurs as a regular and fairly common winter visitor on inland waters. Its stronghold in Devon is Slapton Ley, but it also occurs regularly on Burrator and Hennock reservoirs and Tamar Lake, and irregularly on all other waters.

Pochard are present from October to March, sometimes arriving in August and remaining until April, but the highest numbers, which show great fluctuations and depend on the severity of the weather, occur between November and February. The main numbers recorded at Slapton since 1945 are : 450 in January 1962, 300 in December 1961, 230 in December 1964, over 100 in about seven of these years, and 20-100 in the remainder. Maximum counts for other waters include 115 at Tamar Lake in 1956, 100 at Hennock in December 1961, forty-one at Burrator in February 1959, and forty-five at Wistlandpound in November 1964.

Flocks of up to about fifty occur irregularly on the Exe and Yealm estuaries, and smaller numbers at many other localities. It is seldom reported on the sea, but thirty were observed off the coast near Budleigh Salterton in December 1941. As would be expected, it is rare on Lundy where it has been recorded only twice, singles having occurred in April 1939 and February 1954.

Although there is no proof of breeding in Devon, D & M believed it nested at Slapton in 1895. It appears from their account that the Pochard was less common at that time.

FERRUGINOUS DUCK *Aythya nyroca*

Vagrant

A vagrant from southern and eastern Europe, the Ferruginous Duck was recorded on three occasions during the nineteenth century, namely an undated occurrence at Plymouth and two singles obtained at Slapton Ley in November 1874 and November 1897.

Of the eighteen or nineteen examples recorded during the present century and adequately described, the first was shot on the Kingsbridge estuary on 27 January 1912, BB 5:280. Four occurred at Slapton on 4 December 1949 and the remainder, all singles except where stated, were reported as follows: Tamar Lake January 1952, Exe estuary 31 January 1954, Chivenor 6 October 1954, Erme estuary November/December 1955, the Exe at Topsham 3 March 1956, Tamar Lake 17 April 1957, Fremington 28 September 1958, Slapton 14 November 1959, two at Creedy Pond 29 December 1959, two at Burrator 23 November 1961, and one, possibly the same, 3 January 1962 and Lopwell 17 December 1967.

Those for 1961 onwards were accepted by the Rarities Committee and published in BB with the provision that numbers of these ducks are kept in captivity and some of the records may refer to escaped birds or hybrids. There is no record for Lundy.

GOLDENEYE *Bucephala clangula*

Winter visitor

Although not very numerous, this attractive diving-duck is a regular winter visitor to the estuaries and main inland waters, but is rarely observed on the sea. The principal numbers occur on the Exe estuary, Slapton Ley, and the Kingsbridge estuary, with smaller parties of up to ten, frequently less, on the other estuaries, and occasional birds well up the main rivers. It is regular in small numbers on many of the reservoirs and occasional on practically every other inland water.

The average peak for the Exe during the years from 1944 to 1967 is thirty, with maximum counts of forty-seven in March 1956, forty-four in mid-March 1965, fifty on 19 February 1966, and sixty on 9 February 1967, while the most recorded at Slapton was twenty-five on 8 January 1966 and twenty-four in late February 1954. Twenty-five were observed on the Kingsbridge estuary on 11 February 1956. Adult drakes are always in the minority and represent roughly about a fifth of the total.

In north Devon the numbers are small, usually around four or five on the Taw estuary, but up to ten have been reported at Wistland-pound where it is fairly regular, and smaller parties at Tamar Lake.

Fairly late in arriving, the first individuals are usually seen in about mid-November, but it tends to stay after many other ducks have left, and parties of displaying birds are frequently observed on the Exe during the first half of April. The maximum numbers are recorded from January to March.

The only dated record for Lundy is that of a single bird seen by F. W. Gade on 30 October 1941.

The information given by D & M is not sufficiently detailed to indicate whether there has been any significant change in status during the present century.

LONG-TAILED DUCK   *Clangula hyemalis*

Winter visitor

The Long-tailed Duck, a regular but uncommon winter visitor, occurs annually in ones and twos on the Exe estuary and fairly regularly, if not annually, in Torbay. Its visits to other localities on the south coast are not so frequent, but it occurs irregularly at Slapton Ley, on most of the estuaries, and at times on the sea off Sidmouth and Budleigh Salterton. Although mainly a maritime species during the winter, it also occasionally visits Burrator reservoir and Tamar Lake, and has been reported fairly frequently at Wistlandpound. This diving duck rarely, if ever, occurs on the north coast, and has not been reported from Lundy, but there are occasional records from the Taw estuary and as already stated it sometimes visits the reservoirs in north Devon.

Records involving more than the usual one or two individuals include five at Budleigh Salterton during February and March 1919, five on the Tamar in February 1935, four on the Exe estuary in November 1938, and four at Torquay during March 1962. It is usually well into November or December before the first birds arrive, but they often stay until the end of March and occasionally into April. The comparison of present-day records with those of the nineteenth century suggests that there has been no significant change in the status of this duck.

E

VELVET SCOTER   *Melanitta fusca*

Winter visitor

The Velvet Scoter, almost entirely maritime outside the breeding season, is nowadays a regular though uncommon winter visitor to the south coast, but is scarce and irregular in the north. It seems to have been rather less frequent in D & M 's time, being then regarded as a casual winter visitor.

Probably due to lack of observers, the records for the first four decades of this century are intermittent, but they include a report of sixteen off Budleigh Salterton during most of December 1920 (*BB* 14:235). Since 1938, however, the species has been recorded in all but three of the years to 1967. Although fairly regular, if not quite annual, off the Exe estuary at the present time, the numbers are usually well below a dozen, but a flock of seventeen was present from November 1949 until February 1950. An exceptionally large flock of seventy-five, which did not remain, was seen there by F. R. Smith on 6 October 1956, and up to twelve were present from January to mid-April 1958.

The Velvet Scoter also occurs with some regularity in Torbay and Start Bay, twelve being seen off Slapton in March 1957 and smaller numbers at various other localities during recent years. It has not been reliably recorded from Lundy, but the few records for north Devon include a party of six off Baggy Point on 29 December 1963 in addition to isolated occurrences on the Taw estuary, Peppercombe, Ilfracombe, and Morte Point.

SURF SCOTER   *Melanitta perspicillata*

Vagrant

Of the three scoters on the British list, much the rarest is the Surf Scoter, a North American species which has been recorded five or six times in the county. D & M cite three occurrences, all of birds obtained on the south coast during the latter half of the last century. Of these, the first was shot in Torbay in 1860, the second was killed at Slapton in about 1862, and the third, an immature male, was shot near Kingsbridge on 20 October 1891. On 6 March 1924 a pair was observed off Langstone Rock, Dawlish by W. Walmesley White and others, the occurrence being fully documented in *BB* 17 : 311. A duck considered to be of this species was seen off Teignmouth on 8 January 1929, and is listed in the Third *Devon Report*. The most recent occurrence, and the only one for the island, is contained in the *Lundy Report* for 1956 and relates to a female or immature bird seen in the Landing Bay at Lundy on 16 November 1956.

Since writing this account, a duck was seen by M. R. Edmonds on the Kingsbridge estuary on 15 December 1968.

COMMON SCOTER *Melanitta nigra*

Winter and non-breeding summer visitor

The status of the Common Scoter is difficult to define because its movements are particularly irregular. Although reported annually and in some years occurring in every month, its numbers fluctuate considerably from one year to another. Likewise, it is more plentiful during summer in some years, but in others is commoner in winter.

A strictly maritime duck, away from its breeding grounds, it occurs in compact flocks off the entire south coast, and in the north mainly to the west of Morte Point. Single birds or small parties occasionally occur inside the estuaries, and there are rare inland records for Wistlandpound, Melbury and Hennock reservoirs and elsewhere.

On the north coast during recent years flocks of 100-200 have not infrequently been observed off Morte and Baggy Points, and 300 were seen in Woolacombe Bay during January and February 1947. A flock of 200 was reported off Baggy Point in January 1953 and 250 off Croyde in February 1955. Smaller numbers are fairly regularly noted in Bideford Bay and sometimes at the mouth of the Taw.

Larger flocks occur off the south coast, where the most recorded was about 600 off Dawlish Warren in July 1956. Flocks of about 500 were seen off Dawlish Warren in March 1931, off Dawlish in February 1950, and off Exmouth in July 1958; a flock of 350 was noted off Branscombe in January 1956; and flocks of around 300 off Dawlish in December 1949, and off Brandy Head in December 1962 and February 1964. Smaller numbers of anything up to 200 occur frequently at all points of the coast.

Ones and twos have been recorded at Lundy on about seven

occasions since 1947, and small flocks are occasionally encountered between there and the mainland.

EIDER *Somateria mollissima*
Non-breeding resident

For the entire nineteenth century D & M quoted only about a dozen occurrences, mostly of ones and twos, and the Eider continued to be a rare bird until the 1940s, with no more than a further twelve records for the years 1900-46. In 1947 a party of four occurred on the Exe estuary in November and six on the Taw and Torridge in December, but none were recorded in 1948. From 1949, however, it has occurred annually in fluctuating numbers and in all months of the year, on both coasts but rarely east of Morte Point in the north.

A bird of tidal water, it occurs on the north coast mainly on the Taw and Torridge estuary and off Baggy Point, and on the south coast mainly on the Exe estuary, Torbay, and Start Bay, but small parties have been reported from many localities, and the largest flock recorded, which numbered 150 including at least fifty adult males, occurred at Ladram Bay on 16 November 1958. Although far more frequent in winter than summer, there are many records of summering, non-breeding birds, especially on the north coast.

The principal numbers recorded are: Exe estuary thirteen in February and March 1950, forty-five in March 1954, thirty-five in December 1957, Orcombe Point up to forty-nine in January and February 1963, Taw estuary twenty-five in February 1963 and thirty-two in October 1966, and Baggy Point thirty-seven in February 1963.

The *Lundy Reports* to 1966 contain only two records, both of singles, in the springs of 1963 and 1966.

RED-BREASTED MERGANSER *Mergus serrator*
Winter visitor

A regular winter visitor, the Red-breasted Merganser is nowadays a common species on the Exe estuary, where it has increased very considerably since the latter part of the last century. The earlier writers of the nineteenth century regarded it as an uncommon bird, single occurrences of which were considered worthy of record, but D & M writing at the close of the century referred to it as 'a winter visitor of rather frequent occurrence'.

Despite the numbers that now visit the Exe and to a lesser extent the Teign, it is still an uncommon duck on the other estuaries, and very rarely occurs on inland waters, there being just one record of a single bird seen by G. H. Gush on Paignton reservoir on 22 March 1952. A maritime species, the colourful Merganser frequents the estuaries and their adjacent shallow seas, but is absent from the deep waters around Lundy, most of the north coast, and the more precipitous parts of the south coast. Odd birds and occasional small parties of four or five occur on the estuaries of the Taw and Torridge, Yealm, Dart, Tamar, and the Kingsbridge estuary, with larger numbers of up to thirty or forty on the Teign estuary, where a maximum of eighty-four was recorded on 15 December 1957. Occasional birds are also observed on the sea at various other points along the south coast. The main flocks, however, gather on the Exe estuary and the sea off Dawlish Warren, where counts of up to nearly 200 have been recorded. The average for the 23 years from 1944 to 1966 is 114, with maxima of 196 on 14 December 1947 and at least 180 on 27 November 1955. On the Exe the first arrivals occur as a rule during October, the peak numbers being reached in November and December, usually the latter, but sometimes in January. Most have left by the end of March, but a number of pairs linger on into April, and occasional birds have remained throughout the summer. The sexes in these winter flocks are fairly evenly matched, adult drakes accounting for a little under half of the total.

As already stated, this species is very rare on Lundy where it has been recorded only once, when a single bird visited a small pond on the island, during very cold weather, on 19 December 1938.

GOOSANDER *Mergus merganser*

Winter visitor

More addicted to freshwater than is the previous species, the handsome Goosander is a regular but uncommon winter visitor, except in severe winters when more than usual arrive. It travels far up the larger rivers, is regular on the Dartmoor reservoirs and Slapton Ley, and has been recorded at intervals on every estuary and inland water. The comparison of present-day records with those listed by D & M shows that this species is nowadays more frequent than during the nineteenth century.

In mild winters fewer than five singles may be recorded, but never a year passes without any, while in an average year about fifteen occur, and in severe winters up to about thirty. In the particularly severe weather of early 1963, about 140 were reported from sixteen localities, including thirty-six on the Exe estuary, thirty-five on the Torridge at Landcross, thirteen on Tamar Lake, eleven at Bishops Tawton, and ten at Burrator. The main numbers recorded in other years are : thirteen on the Exe in January 1941 and thirty-two in March 1942, eleven on the Teign estuary in February 1950, thirteen on Slapton Ley in December 1951, and nineteen including nine adult drakes on the Exe in February 1954.

Most of the occurrences are during the months of December to February, but there are records for all months, as the occasional bird has remained throughout the summer. Females and immature birds predominate, but some adult drakes occur, particularly in cold winters.

The only record for Lundy is of a drake on 17 December 1934.

SMEW *Mergus albellus*

Winter visitor

An almost regular winter visitor, the Smew has been recorded in all but four of the thirty years from 1938 to 1967, the exceptions being 1949, 1959, 1961, and 1967. The number occurring in the county is some indication of the severity of the weather, none at all being seen in open winters, but varying numbers of up to almost thirty occurring in cold winters such as 1947, 1954, and 1963, while in February 1956 over seventy were reported in different parts of Devon.

Most birds occur on the south coast, where they are more frequent on the Exe estuary and Slapton Ley than elsewhere, but they do occur at intervals on most of the other estuaries and sometimes on inland waters. In the north of the county, where the Smew is less regular, the records are mainly from the Taw estuary and Tamar Lake. Although the records cover the months from November to

March, most birds occur during January and February, and females and immature birds greatly outnumber adult drakes.

Occurrences of more than the usual numbers include thirteen on the Exe estuary in January 1941, inclusive of four adult drakes; twenty-one, including three drakes, in the same locality in February 1947; five on Horsey Ponds bordering the Taw estuary in March 1947; fifteen, including six drakes, at Slapton in February 1954; twenty-three on the Exe estuary in February 1956; seventeen at Slapton and eleven on the Taw estuary in February 1956; and eight at Clamoak on the Tamar on 18 March 1962. Occurrences at inland waters include very infrequent visits to Burrator, Fernworthy, Creedy Pond, and Shobrooke Park. The only recorded occurrence for Lundy relates to a single bird seen on the early date of 15 September 1933.

The status and records cited by D & M do not suggest that any change has occurred since then.

SHELDUCK   *Tadorna tadorna*

Resident and winter visitor, breeds

After a decrease during the nineteenth century, the Shelduck was already beginning to increase around the 1890s. D & M wrote: 'Resident in very limited numbers but mainly a casual visitor in winter and early spring . . . still breeds on Braunton Burrows'. They added that one pair had nested annually on Dawlish Warren, presumably about the 1890s. The *VCH* subsequently stated that it appeared to be increasing.

The increase has continued throughout the present century and the Shelduck now breeds on every estuary, including the Dart, and also at Wembury and possibly elsewhere on the coast. The continuing increase is apparent from the number of young on the Exe estuary, its main breeding station, where eighty-nine were counted in 1938 compared with 207 in 1964.

The greatest numbers occur from January to March, with a maximum of 603 on the Exe on 11 March 1962, and around 150-200 each on the Taw, Kingsbridge, Plym and Tamar estuaries in recent winters, and considerably smaller numbers elsewhere. The species is practically confined to tidal water and rarely occurs at Slapton or on other freshwater, but has been observed at Tamar Lake on occasion.

Most of the adults apparently migrate to Bridgwater Bay to moult, leaving the estuaries in June and July and flying direct overland. The young meanwhile assemble in flotillas under the care of a pair of adults, which may make the moult migration later. The return to the estuaries takes place from about late October to December. A bird picked up at Dartmouth, however, in March 1963 had been ringed in the Heligoland Bight in July 1958, and it is thought that some Devon birds moult there.

The only occurrences on Lundy refer to a few birds on 25 January 1881 and a single on 28-30 August 1966.

RUDDY SHELDUCK *Tadorna ferruginea*
Vagrant

The Supplement to *The Birds of Devon* records that three examples of this Mediterranean and Asiatic species were shot at the mouth of the Taw estuary in the summer of 1892 when many occurred in different parts of the British Isles. This record, which is not included in *The Handbook* in the list of counties visited by this species in its spectacular immigration of 1892, appears to have been overlooked, due perhaps to its not being in the first edition of D & M's work. Although there is no doubt at all that these were genuine wild birds, the same cannot be said of the other Devonshire records, as the Ruddy Shelduck has for long been kept in captivity in Britain, and isolated occurrences are therefore always suspect. A specimen shot at Braunton in the winter of 1890-91 was considered to have escaped from capitivity.

F. W. Gade records in *BB* 38 : 178 that a bird of this species visited Lundy on 16 September 1944, but the editors of the journal were obviously doubtful of its being a wild bird. There is no really satisfactory record of this species in Devon during the present century.

GREY LAG GOOSE *Anser anser*
Irregular winter visitor

D'Urban considered the Grey Lag Goose to be of no more than casual occurrence, and was able to cite only about seven dated records, mostly of singles, for the entire nineteenth century and none for the first quarter of the twentieth. The position of this species is now confused by the introduction of feral birds, to which several recent out of season records apparently refer. For this reason the few records from 1965 onwards have been ignored.

The *Devon Reports* from 1928 to 1964 list about seventeen occurrences, including the unsubstantiated and rather unlikely record of a skein of seventy-four identified in flight over the Taw estuary in January 1936 which are much more likely to have been White-fronted Geese. The remaining records involve approximately thirty individuals, most of which occurred singly or in pairs, and frequently with small parties or flocks of other grey geese, usually White-fronts. The most recorded was a flock of seven which were closely observed on the ground at Braunton Marsh on 9 November 1947. Parties of four were seen on the Exe estuary on 19 January 1963 and 1 November 1964. Of the others, all of which occurred between the months from November to March, six were reported from Exminster Marshes and the Exe estuary, a pair from the Erme estuary, and one each from the Tamar, Teign, Avon, Axe, and Otter estuaries, and Wistlandpound reservoir.

The two singles recorded on Lundy in September 1949 and April 1964, and one mentioned by Perry, are all considered, because of their tameness, to refer to hand-reared birds.

WHITE-FRONTED GOOSE *Anser albifrons*
Winter visitor

Of the five grey geese, the White-fronted is the only one that occurs at all regularly in Devon, having been recorded in all but about three of the years since 1934, usually in the months from December to February, occasionally early March. It occurs principally in hard winters, the numbers increasing with the severity of the weather, whereas in mild winters very few are recorded.

Most of the records are from the fresh marshes bordering the estuaries of the Exe, Taw, and Tamar, but occasional small flocks are observed on most of the other estuaries, Tawstock Marsh, Wistlandpound reservoir, Tamar Lake, and Slapton. Quite often the birds do not remain long, but in severe winters they may be present in varying numbers from December to March.

The main records from the *Devon Reports* are : Exminster Marshes, 140 in February 1940, 120 in January 1941, 150 in January/February 1945, 100 in February 1951, and 100 in January 1962; Taw estuary, 120 in February/March 1947. Many more than usual occurred in January/March 1963 when some 200 were seen on the Otter, up to 500 on the Exe, 130 at Slapton, 187 on the Tamar, and up to 375 on the Taw.

Ones and twos have been reported very irregularly on Lundy, where a maximum of eight was recorded in October 1949.

One gains the impression from D & M that the visits of this species were more erratic during the latter part of last century, while for the early years of the present century the records are fragmentary.

BEAN GOOSE   *Anser fabalis*

Vagrant

The Bean Goose can nowadays be regarded as no more than a vagrant to Devon, having been reliably recorded on the mainland on only four or five occasions during the present century. D & M considered it to be more frequent than the Pink-footed Goose, but in view of the similarity of the two, and of the confusion that at one time existed between the four grey geese, it is difficult now to assess any change of status that may have occurred during the past hundred years or so. It appears, however, that the Bean Goose, always an uncommon bird in the county, has decreased somewhat during the period under review.

The first record since December 1890, when one was shot on the Kingsbridge estuary, is from Dr F. R. Elliston Wright of one that was shot at Braunton on the Taw estuary on 18 February 1940; the second occurrence concerns a single bird 'very clearly identified' by A. H. Macpherson in the Otter valley on 12 December 1942; while the third, again a single bird, was present on the Exe estuary at Turf from 26 December 1951 until 2 March 1952, during which time it was closely studied by a number of competent observers. During the very severe winter of 1963, two were identified amongst other grey geese on the water meadows at Brampford Speke, by H. Schmalfuss on 17 February, and on 6 March two, possibly the same birds, were recognised by F. R. Smith amongst Pink-footed and many White-fronted Geese on Exminster Marshes.

D & M record that a small flock of Bean Geese visited Lundy in the hard winter of 1860-61. A party of nine was reported there in December 1932, and ten during October 1935. A single bird which visited the island during March/April 1940 was shot.

PINK-FOOTED GOOSE *Anser brachyrhynchus*
Irregular winter visitor

Until 8 February 1895, when one was shot on Aveton Gifford Marsh, there was no satisfactory record of the Pink-footed Goose in Devon. Since then there have been several records from both coasts. D'Urban's MS states that a second was shot at Aveton Gifford in December 1899, following which there is no further record until October 1933 when three were reported at Bideford.

On 25 February and 3 March 1940 seventeen Pink-feet were identified by R. G. Adams amongst a flock of about 120 White-fronts on Exminster Marshes, since when small parties have several times been observed amongst the flocks of White-fronts that visit this marsh in hard weather, usually between the months of December to March. Three were recorded on the Exe estuary on 4 February 1945 and one on 21 December 1946. In 1951 three were present from late January until mid-March and in 1952 one occurred on 6 December. During January 1953 up to three frequented Exminster Marshes in company with two Grey-lags. On the Taw estuary one was observed during February 1955 in a flock of White-fronted Geese.

In the cold winter of 1963 one occurred on the Axe in January, six or seven on the Exe in February and March, and two at Slapton, three on the Plym, and two at Brampford Speke during February. On 7 December of that year eight were seen on the Exe Marshes, and one occurred at Tamar Lake from about 20 December until early February 1964.

The few records of this species on Lundy relate to two on 24 January 1940, one in December 1949, one on 1 October 1959, and one in October and two from 2 to 9 November 1960.

BRENT GOOSE   *Branta bernicla*

Winter visitor

The Brent Goose, a regular winter visitor to the Exe estuary from about November to mid-March, is now fortunately increasing in numbers, under protection, after a serious decline which reached its lowest level during the early 1950s. The race concerned is the Dark-breasted form *(B.b.bernicla)* although a few individuals of the Pale-breasted form *(B.b.hrota)* are recorded most years, including a flock of twenty-four passage birds noted on 19 April 1962.

Up to about 1940 the species occurred not only on the Exe but also on the Taw and Torridge, the Otter, Tamar, and occasionally at Slapton and elsewhere, but is now rare away from the Exe.

D & M stated that this goose occurred in flocks of considerable size in our bays and estuaries, but omitted to give any indication of the numbers, except that a flock of about 120 frequented the Exe below Topsham for ten days in February 1855. W. Walmesley White reported 'great numbers' on this estuary in January and February 1917, and February 1929. Rev F. L. Blathwayt recorded a flock of about 400 off Dawlish Warren on 4 May 1932, evidently passage birds.

Although the maximum number wintering in 1932 was recorded as only seventeen, it was given as 300-400 on 25 January and at least 250 at the end of February 1933. During the remainder of the 1930s the number of wintering birds fluctuated from sixty-three in early 1935 to 272 in January 1939, and 200 on passage on 26 April 1936. It gradually decreased during the 1940s from about 350 in January 1941 to thirty-four in February 1945, ninety-nine in February 1947, and only twenty-eight in January 1949. From 1950 to 1955 it remained around thirty, but from 1956 onwards, following protection, the numbers slowly increased to eighty in February 1958. During the 1960s the increase has continued but fluctuates markedly according to the success of the previous breeding season. In 1961 the maximum was 111 on 12 February, but only forty-six in January 1962, followed by 118 in February 1963, when unusual numbers were seen in other localities during the severe winter.

In 1967, after a very good breeding season in the previous year, 170 were counted on 25 January, of which about a third were young.

Flocks of passage birds, noted often on the Exe, have also been recorded at Slapton, where thirty-five were seen on 29 April 1934 and twenty-one on 25 April 1964. Singles have been observed on Lundy, at irregular dates, on about five occasions since 1933.

BARNACLE GOOSE *Branta leucopsis*

Irregular visitor

The Barnacle Goose, which until recently had sadly decreased as a winter visitor elsewhere in the British Isles, seems never to have been more than a vagrant or irregular visitor to Devon. D & M knew of only about eight or nine occurrences, mostly of single birds or pairs, during the whole of the nineteenth century; the only considerable number was that of a large flock which visited Slapton Ley in the winter of 1801, and is recorded in Montagu's *Supplement*.

The nine or ten records for the present century are of much the same pattern and comprise mostly single birds, and a few small flocks. The first relates to one which W. Walmesley White saw on the Otter estuary on 30 March 1932 and reported in *BB* 25:363. The next two occurrences are of single birds seen on the Taw estuary at Instow on 20 October 1934 and at Dawlish Warren on 30 January 1935. On 5 December 1959 S. G. Madge watched a single adult with a flock of Canada Geese at Shobrooke Park. It was seen to leave during the afternoon, and within half an hour was observed by F. R. Smith arriving on the Exe estuary marshes, where it remained with a number of Canada Geese until 14 December. During this period it was frequently seen by R. G. Adams, P. W. Ellicott, and myself, and the incident is recorded in the *Devon Report* for 1959. The most interesting recent record, however, is that of a flock of thirty-one which D. B. Cabot witnessed coming in from the sea at Dawlish Warren on 2 January 1962. They passed within 15 yd of him and continued north-westwards up the Exe estuary. They were followed two days later by two other birds on the same course, as related in the 1962 *Devon Report*.

There have been four Lundy records during the past thirty years: a single bird that remained on the island for almost a month from 24 April 1941, and a party of three on 6 September 1944, both recorded by F. W. Gade; one from 26 April to 5 May 1959; and a party of six which were present from 29 October to 2 November 1966—*Lundy Reports* for 1959-60 and 1965-6.

CANADA GOOSE   *Branta canadensis*

Resident, breeds

The Canada Goose, an introduced species, was listed by D & M, who referred to it as: 'Introduced but has occurred several times . . . apparently in a wild state'. Until 1949, when the species was introduced at Shobrooke Park by the owner, Sir John Shelley, the *Devon Reports* contained only three records of this bird, of which the first referred to a single one seen on the Exe estuary in January 1940.

From the original ten young reared at Shobrooke Park, the number had increased to at least ninety by 1959, and the birds were already spreading to other localities. A pair bred unsuccessfully at Bicton Lake in 1956, and in the following year birds were reported as far apart as the Exe estuary and Fernworthy reservoir, and were breeding at Fulford Estate. By 1966 they had been breeding intermittently at a number of places including North Tawton, Bicton, and Burrator, and were occurring in many parts of the county. The *Devon Report* for 1967 includes counts of 203 at Shobrooke on 12 January, and 100 on the Exe estuary on 13 August, but there is much movement between different localities, and a favoured feeding place is the flood waters of the Exe at Brampford Speke where 187 were counted on 1 January.

RED-BREASTED GOOSE   *Branta ruficollis*

Vagrant

Two very old Devon records of this Siberian species are listed by D & M; the first relates to a bird that was killed at Dawlish Warren in 1828, and the second to one shot on the Teign estuary on 21 February 1837.

There is a single occurrence for the present century, relating to a bird that accompanied a flock of 150 White-fronted Geese at Bickham during the very severe winter of 1963. It was present from 4 to 8 March and was first seen and identified by A. MacFarlane and subsequently observed by Lord and Lady Roborough, Tony Soper, and others. This record was accepted by the Rarities Committee, in *BB* 57:266, and the bird is fully described in the *Devon Report* for 1963.

MUTE SWAN *Cygnus olor*

Resident, breeds

This species has greatly increased since D & M's time, for they used to see only about twenty-five on the Exe at Topsham up to around 1856, after which the swans were reduced by the salmon fishermen to only a few pairs. D'Urban noted that a few pairs were kept on the Exe and on many ponds and ornamental waters throughout the county.

A census organised for the BTO by Bruce Campbell (*Bird Study* 7:211) gave the total number in 1955 as 371 birds, including fifty-seven breeding pairs, and in 1956 there were 406, including forty-eight breeding pairs. A repeat census in 1961 gave a total of 597, inclusive of seventy-eight breeding pairs (*Bird Study* 10:14). The breeding pairs are dispersed throughout the county on ponds, lakes, the lower reaches of rivers, and creeks, but none nests on the Dartmoor reservoirs where they rarely if ever occur.

Most of the non-breeding birds are confined to the Exe estuary where the numbers have fluctuated, gradually increasing to a maximum of 302 birds in September 1961, but since decreasing to around 150 during the past few years, the peak numbers usually being recorded in August or September. Herds of 30-90 occur less regularly on most of the estuaries and at Slapton, but there is considerable movement from one locality to another during the course of a year.

The only record for Lundy refers to a single bird on the sea on 10 September 1959.

WHOOPER SWAN *Cygnus cygnus*

Winter visitor

D & M considered the Whooper Swan to be much rarer than Bewick's and to occur only in severe winters. Except for five at Braunton

Marsh in December 1906, there are no records until 1941, but from then onwards it was reported in 21 of the 27 years to 1967, involving about 160 birds, nearly half of which occurred in small parties of up to four or five in widely scattered areas including most of the estuaries and reservoirs. The maximum numbers recorded are nineteen on the Tamar on 9 November 1952, ten on the Exe estuary on the same day, and fourteen on the Exe on 11 February 1956.

In the severe winter of early 1963 about twenty were reported, including four each at Slapton Ley and the Axe estuary, and seven at Stafford Bridge on the River Exe. Most of the occurrences have been in the months of November to February, occasionally in October, and wintering birds have sometimes remained until March and even early April. Parties of up to four or five have wintered on Tamar Lake in six or seven winters during this period and the species has occurred several times on the Taw and Torridge estuary.

The total includes two records for Lundy, the first of six birds which stayed for a fortnight during November 1949, and the second of two which remained for a few weeks in November and December 1962.

BEWICK'S SWAN   *Cygnus bewickii*
Winter visitor

D & M knew of only three or four occurrences during the whole of the last century, including a flock of fifteen on the Exe in January 1838. Other than the record in the *VCH* of nine on the Kingsbridge estuary in December 1902, there are none for the present century until 1930. Between then and 1960 about fifty, mostly singles, were recorded in the county, including a flock of thirteen on Horsey Pond, Braunton, in January 1947 (which were recorded as Whoopers when seen flying over Weare Giffard on 26 January), and four each on Beesands Ley in March 1956 and the Exe estuary in January 1960.

From 1961 the species has become much more frequent and about 150 were recorded in the seven years to 1967, including nearly fifty in January 1962, at least fifty in the severe winter of early 1963, and fifteen in 1964. The largest numbers recorded are twenty-three on the Axe estuary in January 1962, fourteen on the Exe in January 1962, fifteen on the Axe in January 1963, nine on the Plym in December 1963, fourteen on the Axe in January 1964, eight on the

Page 85:
Montagu's Harrier,
female with young.
A very rare summer
visitor, one or two
pairs of which
have nested almost
annually during
the past fifty years

Page 86:
The Erme estuary.
One of several
sheltered estuaries
on the south coast,
it is visited by
many different
species during the
course of a year

Exe in January 1966, and six at Slapton in January 1967.

Bewick's Swans have been recorded many times on Tamar Lake where six were seen in November 1962 and six from January to March 1963. Most of the occurrences are in the months from December to February, and in the past were mostly in severe weather. The other records are from widely separated localities including the Taw estuary, the Exe at Stoke Canon, the Culm where eight were recorded in January 1955 and seven in January 1962, the Otter, and Burrator reservoir. In most cases the birds have remained for only a few days, but they have occasionally wintered. There is no record of this species from Lundy.

BUZZARD *Buteo buteo*

Resident, breeds

A familiar sight everywhere in Devon, the Buzzard is widely distributed throughout the county, and is only slightly less common than the Kestrel. While rabbits were still plentiful, it bred commonly along the cliffs of both coasts but does not do so at present. Inland it breeds plentifully both in wooded and agricultural country, in the conifer plantations and in scattered trees on Dartmoor and Exmoor.

D & M described it as resident and not uncommon, but constantly decreasing and confined to the coastal cliffs and wilder parts of the county. In the *VCH* D'Urban stated that it still nested annually in several localities, while his subsequent notes refer to an increase during the 1910s and he mentions that seven pairs were observed in the Teign valley above Steps Bridge in April 1911. In 1929 Loyd reported that it was increasing as a breeding bird in the south-east, where he knew of many pairs.

An investigation in 1929 by H. G. Hurrell and V. C. Wynne-Edwards put the number of birds at 900-1,200 and concluded that it was more plentiful in the north than the south. Sample counts in four separate areas in 1954 gave a population of fifty-nine breeding pairs in about 52 square miles, at which date it had probably reached its maximum density. In 1952 G. H. Gush counted twenty-nine in the air together between Ilfracombe and Georgeham on 19 March, and the same number at Eggesford a fortnight later.

Following the advent and spread of myxomatosis in 1954, which all but exterminated the rabbit, few pairs of Buzzards bred success-

F

fully in 1955. P. J. Dare, who made a close study of this species on Dartmoor during the 1950s, wrote in *Countryside* 18:306-314: 'In Devon, immediately before myxomatosis, parties of up to thirty Buzzards could be seen soaring high in the spring skies . . . while in one south Devon valley, an area of only eight square miles supported no less than twenty-one breeding pairs of Buzzards. The normal density for this species is a little over one pair per square mile'. He stated that in the worst affected areas up to a third of the population came to grief.

After a period of gradual recovery, the numbers were again partially reduced by the severe winter of 1962-3, followed by further recovery since. In 1968 P. J. Dare reported that the population in the Postbridge area of Dartmoor had been stable since 1960, apart from the temporary drop in 1963, and varied between ten and fifteen breeding pairs in an area of fifteen square miles.

The total population is considered to be not less than 1,000 pairs at the present time.

Although the Devon population is almost entirely sedentary, L. H. Hurrell noted a movement of fourteen flying northwards up the Avon on 11 April 1955.

On Lundy, where it is also resident, the number of breeding pairs was recorded as five in 1939, probably four in 1947, two or three in the years up to 1952, one in 1953 and 1954, followed by unsuccessful attempts by one pair in most of the years to 1965, while in 1966 only a single bird remained. Although other birds are occasionally reported, there appears to be no evidence of regular migration at Lundy.

ROUGH-LEGGED BUZZARD   *Buteo lagopus*

Vagrant

Not all the sight-records of the Rough-legged Buzzard can be relied upon owing to the variable plumage of the common Buzzard, light examples of which appear at times to have been mistakenly identified as the rarer bird. Specimens obtained during the nineteenth century, however, have conclusively proved that this migrant from northern Europe does occasionally reach Devon during the winter months.

D & M cite ten occurrences of birds obtained between the years from 1836 to 1883, including four from Dartmoor, two from the fringes of Exmoor, and one from Braunton Burrows. Three of these were trapped in north Devon during the winter of 1875-6, when the species occurred in unusual numbers in Britain. The Ilfracombe Museum, according to the *Ilfracombe Fauna & Flora*, contains three specimens 'that were obtained locally many years ago'. Of the nine records in the *Devon Reports*, some bear insufficient evidence of identification and may well have referred to the common species, while others are unconvincing. The example recorded by W. J. Wallis on the Dart estuary during the first half of February 1948, however, and further described in vol 2 p 11 of *Devon Birds* appears to be authentic, as does the bird observed by Lady Drewe at Clyst Hydon between 28 November 1948 and 14 February 1949. The most recent record of one observed on Haldon on 28 and 29 May 1967 is fully documented. In addition to these, a note by W. Walmesley White in *BB* 13:314 gives details of one that he observed at Budleigh Salterton on 29 March 1920. A further record, mentioned in the *Ilfracombe Fauna & Flora*, concerns a bird that Richard Perry watched near Georgeham at the end of January 1939, which he described to me as having noticeably feathered tarsi that were conspicuous as the bird moved about on the ground.

The *Lundy Report* for 1959-60 records that a Rough-legged Buzzard occurred on the island on 29 and 30 November 1959. It was observed closely in Millcombe on 29 November, and on the following day was seen to fly south.

SPARROW HAWK   *Accipiter nisus*

Resident and passage migrant, breeds

D'Urban considered this to be the commonest hawk in Devon, despite the persecution it then suffered from game-keepers. Although it is now vastly outnumbered by both the Kestrel and Buzzard, it is holding its own, after a decrease during the late 1950s due apparently to toxic chemicals.

In the 1930s and 1940s it was quite common, and in 1946 S. D. Gibbard reported that an average of 50-60 were killed every year on one estate alone, while in 1951 H. G. Hurrell knew of at least six nests within a 3-mile radius of his house on the southern edge of Dartmoor. The numbers seem to have been at their lowest around

1960, since when they have increased from being recorded in thirty-three localities in 1962 to about ninety in 1966 and over 100 localities in 1967. It appears from the *Devon Report* that in 1967 a pair was resident in almost every 10 km square of the county.

The Sparrow Hawk was recorded by Loyd as breeding on Lundy in 1922. In the years from 1947 to 1963 a few singles were noted annually on spring or autumn passage, with a maximum of thirteen records in 1949, only one each in the years from 1958 to 1963, and none since.

GOSHAWK   *Accipiter gentilis*
Vagrant

The Goshawk was not admitted to the Devon list by D & M, who considered that the one or two reported occurrences were extremely doubtful. The *Devon Reports* contain fully detailed accounts of nine occurrences, all of single birds:

1932   Noss Mayo, 3 April (G. M. Spooner)
1945   Roundswell, Barnstaple, seen twice during January
         (H. S. Joyce)
1952   Spreacombe, 13 December (R. Newton)
1957   Postbridge, Dartmoor, 21 March (P. J. Dare)
1957   Spreacombe, 28 and 29 December (D. Wilson &
         F. R. E. Wright)
1959   Spreacombe, 13 September (D. Wilson)
1960   Puslinch, near Yealmpton, 12 May (L. I. Hamilton)
1960   Exmoor, on the Somerset border, near Pinkworthy Pond,
         15 September (C. J. F. Coombs)
1967   Stoke Point, near Newton Ferrers, 7 October (R. Burridge)

The only reliable record for Lundy, which concerned a first-year bird seen on 17 and 18 April 1951, was square bracketed, as being possibly an escape from captivity.

RED KITE   *Milvus milvus*
Irregular visitor, has bred

An irregular visitor occurring in all months but most frequently in spring, the Red Kite has nested at least once since 1900 and has wintered during the early 1960s. Although said to have been numerous in Devon at the close of the eighteenth century, it had become very scarce by Montagu's time, as he saw only one during his 12 years' residence in the county. Pidsley quotes E. H. Rodd as stating that it bred regularly near Holne Chase around the 1820s, but D & M considered it was never very common and that many old records referred to harriers and buzzards. They gave details of about twenty occurrences during the entire nineteenth century, mostly of birds shot for private collections, but they were not aware of any definite breeding records.

A long account of this species, including its possible breeding on the Yealm estuary around 1885 and unconfirmed breeding near Kingsbridge in 1947, is given by H. G. Hurrell in *Devon Birds*, vol 3. The nesting of a pair in Devon in 1913, referred to in *The Handbook*, occurred near Dartmeet where, according to D'Urban's MS, a pair also bred in 1912.

Of over twenty records of the Red Kite contained in the *Devon Reports* for the years 1932-67, at least half refer to Dartmoor, where singles have several times been observed in the months from March to June, while one was recorded on west Dartmoor during November and December 1962, from March to September 1963, the whole of 1964 (with two on several occasions), and up to March 1965. The latest record refers to one seen near Hessary Tor in March 1967. In addition to the Dartmoor records, the species has been observed at a number of other inland localities, including south-west Exmoor, Lapford, Yarcombe, Sutcombe, Stoke Rivers, and Ashprington. The only dated record for Lundy relates to three birds seen flying together over the island on 28 April 1929 by F. W. Gade.

Records in sixteen of the years since 1940 suggest that this species occurs more frequently in Devon than in any other county outside Wales.

WHITE-TAILED EAGLE   *Haliaetus albicilla*

Vagrant

Seven or eight occurrences during the nineteenth century are enumerated by D & M, who also state that it was most likely this species which at one time nested at the Dewerstone on the River Plym. This eagle is also believed formerly to have bred on Lundy, and an example in the Ilfracombe Museum was shot on the island in about

1880. D'Urban's MS notes refer to one shot at Torcross in November 1899; another was killed at West Buckland in March 1909, and an almost adult bird was shot on the Kingsbridge estuary in February 1909.

Two birds, thought to be this species, were seen near Hunter's Inn on 13 March 1932, *BB* 25:362. The *Devon Reports* contain six records, the most detailed relating to one seen on a number of occasions by G. M. Spooner and others in the Princetown area of Dartmoor during February and March 1936. Another, or possibly the same bird, was observed in this part of Dartmoor on three occasions during June 1938. The latest record relates to a bird seen at Hope's Nose, Torquay on 14 January 1946.

HONEY BUZZARD   *Pernis apivorus*

Vagrant

Describing the Honey Buzzard as a casual visitor of very rare occurrence during the summer months, D & M were able to cite records of only seven birds obtained in the county during the entire nineteenth century. Most of these were shot in south Devon, but one is recorded as having been trapped on Dartmoor in 1848 and another was obtained in north Devon in 1866. Although the Honey Buzzard breeds in very small numbers in southern England, it has been recorded only twice in Devon during the present century. The first was a melanistic male which was shot on Dartmoor on 20 September 1904 and sent to E. A. S. Elliot of Kingsbridge, and is mentioned in *BB* 1:265. The second, a sight-record, is of a bird which M. Huxtable watched flying straight over Fremington Pill on the Taw estuary on 4 May 1958. His description of the bird in the *Devon Report* for 1958 gives the characteristic features of this species, as seen from below.

Since this account was written, a single bird was observed by L. H. Hurrell on the Erme estuary on 11 July 1968.

MARSH HARRIER   *Circus aeruginosus*

Irregular visitor

In their account of this species, D & M wrote: 'The Marsh Harrier, once very common, is now by far the scarcest of the three British

Harriers in Devonshire. On Dartmoor, where it formerly nested, we never encountered a single example in the course of many years' snipe shooting'. Their assessment of its status still holds good, for this fine hawk, an easy prey to gunners, is nowadays no more than a casual visitor to Devon, where it has been reliably recorded on about fifteen occasions during the present century. D & M were able to cite only five occurrences from 1857 to 1890, one of which refers to a pair that frequented the Axe valley near Musbury during the greater part of 1876, and may possibly have bred there.

Dr Edward Moore, who in 1830 produced the first systematic list of the birds of Devon, wrote this of the 'Moor Buzzard': 'I am informed by the warreners on Dartmoor that it is not uncommon and commits great depredations among their rabbits'. Although he differentiated between this species and the Buzzard, which he said was very common, one feels that there was considerable confusion between these two hawks on Dartmoor.

As will be seen from the records that follow, the Marsh Harrier has occurred in practically all months of the year, but more frequently during the autumn than at other times. A few of the records published in the early numbers of the *Devon Reports* have been ignored because no details of identification were given; all those quoted are detailed in the *Reports* for the years stated.

1928 near Budleigh Salterton, a male on 9 July, *BB* 22:64
1930 north-west Devon, a male on 23 May
1944 locality not stated, a male on 23 September (N. V. Allen)
1944 Lundy, two on 15 October and one on 1 November
      (N. V. Allen & F. W. Gade)
1945 locality not stated, a female or immature from 1-5 March
      and one on 9-10 November (R. B. Phare)
1947 Broadsands, two on 1 February (C. E. Hicks)
1947 Braunton Marsh, an adult male on 5 October (R. F. Moore)
1950 near Widworthy, east Devon, a male on 7 February
      (F. C. Butters)
1953 Exminster Marshes, one, probably immature, on
      9 September (J. R. Brock & J. L. Bradbeer)
1958 Slapton Ley, an immature on 1 January (J. J. Hatch), and
      1 February to 25 March (R. M. Curber & M. R. Edmonds)
1958 Lundy, one in Pondsbury area on 30 April (Mary Squires)
1961 Slapton Ley, an adult female on 14 December (E. A. Roberts)

1965   Thurlestone, an immature on 21 October
(T. Soper & J. R. Brock)

1965   Slapton Ley, probably the same individual on 23 October
(R. F. Moore & P. J. Dare)

HEN HARRIER   *Circus cyaneus*

Winter visitor, formerly bred

D & M wrote of the Hen Harrier: 'Still perhaps a resident; but now for the most part only a casual visitor, principally in the autumn, and of rare occurrence'. The only dated breeding record which they mentioned was in 1805, but D'Urban in the *VCH* stated that a female and her four eggs were obtained at Torhill near Throwleigh in about 1861, which was probably the last record of breeding in the county. He appears to have overlooked his *Supplement* to *The Birds of Devon* which states, under Buzzard, that the eggs of the latter and also of the Hen Harrier were taken close to Ilfracombe during the summer of 1893.

There appears to be no fully substantiated records of breeding in Devon during the present century, but Loyd stated that he knew of more than one breeding pair in the county, near Dulverton, but he omitted all details. *The Handbook* mentions the possibility of its having nested in Devon since 1900.

The *Devon Reports* contain about a hundred records of Hen Harriers, but it seems that some of the summer records could well refer to Montagu's. Similarly, a record in *BB* 19 : 180 of a Hen Harrier on Welsford Moor on 11 July 1925 is more suggestive of a Montagu's Harrier, as the latter bred in this locality during the 1930s and probably earlier. In particular, there is a record of a pair, thought to be Hen Harriers, which were present during the whole of 1942 and 1943 and nested successfully in both years, but the locality was suppressed and no details of identification were given.

As a winter visitor, the Hen Harrier occurs regularly, probably annually, on Dartmoor and Exmoor, irregularly on Braunton Burrows and the Taw estuary, Tamar Lake and Haldon, and infrequently at many other localities. The numbers recorded are small, usually two or three singles during the course of a year, but during 1966 and 1967 there were some twenty records for the mainland, including a pair on Molland Common in March, and singles from a number of widely separated areas in the months from October to April. The indications are that this species is now becoming more frequent as a winter visitor.

Davis records four occurrences for Lundy up to 1954. Since then a 'ring-tail' was reported in October 1958, a male and female in November 1963, a female in May 1964, and a 'ring-tail' in March and a male in October 1966.

MONTAGU'S HARRIER   *Circus pygargus*
Scarce summer visitor, breeds

Montagu's Harrier, most beautiful of the three species that occur in Devon, is named after Colonel George Montagu of Kingsbridge, who in 1802 showed it to be distinct from the Hen Harrier and named it the Ash-coloured Falcon. It seems never to have been plentiful, for D & M referred to it as a casual visitor which had been known to breed on several occasions, and of which about thirty examples were obtained in the county during the nineteenth century for private collections. As a breeding species it evidently occurred in a number of widely separated areas, mostly on gorse- and heather-clad commons, and was not confined to Dartmoor.

D'Urban's MS contains no breeding records for the first quarter of the present century, although undoubtedly the species nested at least irregularly, and for this period Loyd was unable to add any fresh information for east Devon. The *Devon Reports*, however, indicate that breeding has occurred on at least fifty occasions since 1928, and that one or more pairs nested in at least 31 of these 40 years, having certainly been assisted and probably saved by the afforestation of moorland during this period. Even so, breeding occurred regularly during the 1930s in certain of the upland marshes of west Devon. The peak numbers, however, occurred in the young conifer plantations during the late 1940s and early 1950s, with thirteen young successfully reared from four nests at one locality in 1950.

It is unfortunately still necessary to suppress exact localities, but it may be said that the species has nested in a number of widely separated parts of Devon during these years, although fewer pairs have bred during the 1960s.

On migration Montagu's Harrier still occurs regularly, and singles or occasional pairs are reported annually from inland and coastal localities during the months of April to September. Of birds ringed as young on Dartmoor, at least four have been recovered, shot, in different parts of France—one in September 1958 a month after being ringed; one in August 1960 when seven years old; one in September 1961, two years old; and one in June 1958 after eight years.

Apart from old records there are eight dated occurrences of single birds on Lundy since 1937 for the months of April to August. The species breeds regularly in Cornwall but not now in Somerset.

OSPREY *Pandion haliaetus*

Scarce passage migrant, formerly bred

Because of its great interest, the account of this species is quoted in full from the *VCH* published in 1906 : 'No example of this fine bird appears to have occurred since the autumn of 1875, when quite a flight seems to have visited Devonshire, and specimens were obtained on the Teign, Dart, Avon, Tamar and Taw. Up to the middle of the last century the Osprey was well known on the estuaries of the larger rivers, especially in the south of the county, being seen at all times of the year, but generally in spring and autumn. It is stated by Polwhele to have bred on a pinnacle of the cliffs at Beer in the eighteenth century, and also on the cliffs of the north Devon coast. There was an eyrie at Gannet's Combe on Lundy Island, where a pair bred as lately as 1838, in which year the male was shot and the female disappeared never to return'.

Except for one shot at Bickleigh on the Plym in September 1905, there is no further record of the Osprey until 1920. From that year onwards the records are too numerous to be quoted individually but they comprise about forty-seven occurrences, of which fourteen were on the Exe estuary, six on the Taw estuary, three on the Yealm, four on the Dart, three on the Axe, two on the Erme, two at Fernworthy reservoir, and two on other parts of Dartmoor. The remaining eleven occurred at various localities along the north and south coasts and estuaries. Of these records, nine were in the month of August, fourteen in September, five each in October and November, four in May, and the remainder at various other times of the year. In 1944 a bird was present on the coast at Ilfracombe from August until early December.

That this magnificent fish-hawk is regaining some of its lost ground may be judged by the fact that twenty-five of the records have occurred during the past ten years, the Osprey having been observed annually on autumn migration since 1958. It has not been seen on Lundy, however, during the present century.

HOBBY   *Falco subbuteo*
Summer visitor, breeds

The elusive Hobby, regarded by D & M as a scarce summer migrant which occasionally bred in the woods on the fringe of Dartmoor, has increased during the past 20 years or so and has bred regularly since about 1950 or earlier. In fact, the *Devon Report* for 1967 stated that there were at least six pairs in the county, of which three were known to have bred successfully and one was robbed.

Although Loyd knew of no breeding records for east Devon during the first 30 years of this century, E. G. Weldon informed W. Walmesley White that a pair nested at Tracey near Honiton in 1937, while the *Devon Report* for 1944 stated that they bred in that area again in 1942 and 1943. In north Devon Dr F. R. Elliston Wright recorded breeding around 1930 and near Braunton in 1946, and B. G. Lampard-Vachell considered that a pair bred near Weare Giffard in 1943.

In another locality previously suspected breeding was first proved in 1951 and again in 1957 and 1959, while it was suspected in a further locality during the late 1950s. By 1961 it was known to be successful at two sites, and in 1964 at three sites, since when the number has gradually increased, although it has not always been possible to obtain definite proof of breeding because of the elusiveness of these birds and the great area of suitable country. For some years the Hobby has also been suspected of breeding on Dartmoor.

Although not mentioned in one or two of the earlier *Reports*, the Hobby has been recorded annually from widely separated parts of the county since 1933, and from Lundy in eight of the years since 1948, in addition to a few earlier records. The occurrences mostly fall between the months from May to September, as the species is one of the last to arrive in the spring, but there is a Lundy record for the end of March and a mainland one for October.

It appears from *The Birds of Somerset* that this beautiful little falcon is also becoming better established in that county.

PEREGRINE   *Falco peregrinus*

Resident, breeds

During the first half of the present century one could expect with confidence to watch these falcons in many parts of Devon, either by visiting the headlands in the summer, or in the winter the estuaries where they regularly harried the flocks of plover and waders.

In the *VCH* D'Urban wrote that the Peregrine was 'a well known resident, not uncommon, and rather on the increase, though subject to increasing persecution. . . . There are many eyries known both on the north and south coasts'.

Until the 1939-45 war, the Peregrine was well distributed in the county, being frequently reported from the coast and inland, and breeding regularly on suitable cliffs along both coasts. There, despite continuous persecution from pigeon fanciers and the toll of young taken by falconers, it managed to hold its own and occupied perhaps between twenty and thirty eyries altogether. During the war many of these fine falcons were killed by the Air Ministry because they constituted a threat to carrier pigeons, but after the war there was a partial recovery in numbers until the mid-1950s when a serious decline started which all but exterminated the entire breeding stock.

The sudden and rapid decline of the Peregrine was subsequently found to be due to toxic chemicals which accumulated in the predators from sub-lethal doses in their prey, causing egg-breaking, sterility, and ultimately the disappearance of the adults (D. A. Ratcliffe, *Bird Study* 10 : 56).

The *Devon Reports* from 1928 to 1957 contain very many records of sightings in all parts of the county, including Dartmoor, to which it is principally a winter visitor. Unfortunately, however, it was necessary to suppress most of the breeding records, and those actually mentioned represent only a part of the total number, which makes it difficult now to assess the total breeding strength during these years. In north Devon, for instance, I watched a particular eyrie regularly from 1933 to 1939 and again for a few years after the war, during which period it was continuously occupied, though breeding was not always successful and on occasions the eyasses when just on the point of flying were taken for falconry. For south Devon, to quote another record, G. M. Spooner states in the *Report* for 1931 that falcons were observed at five eyries in the Plymouth district alone in that year.

The Report for 1958 laments that the species had become much less plentiful but nested as usual in several localities, whilst those for 1959 and 1960 referred to a continuing decrease in numbers. In 1961 only one pair bred successfully and a second pair attempted to breed, both on the north coast. By 1962 only one pair was known to have bred out of twelve eyries occupied in 1955. Since then, however, there has fortunately been a slight recovery, with possibly two pairs breeding in 1965, at least one in 1966, and at least three pairs in 1967.

On Lundy, which was once famous for its falcons, one and sometimes two pairs bred regularly if not annually up to 1938, following which a pair bred successfully in 1950, 1953, and 1955. Since then, although single birds and occasionally two have been recorded annually, there has been no further proof of breeding up to the time of writing.

GYR FALCON *Falco rusticolus*

Vagrant

Of the three races of the Gyr Falcon occurring in Europe, only the migratory Greenland Falcon (*F.r.candicans*) has reached Devon, where some six examples have been obtained or seen during the past 150 years. The first record, about which there is some doubt as to the exact locality, relates to a bird taken on the border of Devon and Cornwall, on the Tamar or the Lynher river, on 7 February 1834. Next there is a specimen, now in the Bristol Museum, which was shot by a Mr Philip Wathen when woodcock shooting on Lundy one November about the middle of the nineteenth century. A third example was reported to have been observed at close range on the cliffs at Rousdon by Henry Swaysland in June 1882. Loyd refers to a supposed Greenland Falcon recorded in the *Field* of 1884 as having been seen at Sidmouth, but rejects the record.

An occurrence of which there can be no doubt at all is that of a very fine male, preserved in the Exeter Museum, which was shot on Lundy on 13 March 1903. Dr F. R. Elliston Wright records that one was taken in a rabbit trap on Braunton Burrows in 1925. Lastly, *BB* 31:92 relates: 'Mr F. W. Gade informs us that a Greenland Falcon visited Lundy in March 1937. The bird was a tiercel, and remained for about three weeks. It was badly mobbed by gulls and crows and seemed to be in rather poor condition. It suddenly disappeared and subsequently its skeleton was found amongst boulders on one of the island's slopes'.

MERLIN   *Falco columbarius*

Resident and winter visitor, breeds

D & M regarded this small falcon as an infrequent autumn and winter visitor, but knew of no definite breeding records. Later D'Urban stated in the *VCH* that it was believed to have nested on Exmoor, and his MS records that a nest was found near Badgworthy Water in May 1907. W. Walmesley White reported in *BB* 15:45 having found a nest on the Devon side of Exmoor in June 1921, and later stated that it also bred there in 1922. In *BB* 28:87 he recorded breeding on Dartmoor in May 1934 and considered that the Merlin had nested in this locality for many years. F. H. Lancum reported breeding on south-west Dartmoor in May 1920 (*BB* 28:120) in a locality where nesting has since been reported.

In north Devon Dr Elliston Wright recorded in the *Devon Report* that a pair nested on Braunton Burrows in 1935, and the 1940 *Report* contained his photographs of a nest with young, taken in that year.

Breeding by one to three pairs is known to have occurred on Dartmoor almost annually, and is suspected in other years since 1953, and probably earlier. On Devon Exmoor breeding was almost certain in 1957, and was proved in 1958, 1964, and 1965.

As a winter visitor, singles are reported annually from widely scattered areas, but most often from coastal localities, and particularly from the Taw estuary, from about September to March. On Lundy it occurs regularly in small numbers, on spring or autumn passage, often both, and has been recorded every year since 1947, with three birds present on 20 October 1958.

RED-FOOTED FALCON   *Falco vespertinus*

Vagrant

This beautiful little falcon, a vagrant from south-eastern Europe, is described by D & M as 'an occasional visitor of extremely rare occurrence'. D'Urban in the *VCH* lists three examples said to have been obtained and a fourth seen in the county, but admits that none of these records is properly authenticated.

Until quite recently the only record for the present century, though a somewhat doubtful one, was that contained in Loyd's *BSED* of a supposed Red-footed Falcon seen by C. A. Smith at Sidmouth in June 1927 and rather inadequately described in a letter quoted by Loyd. The *Devon Report* for 1957, however, gives satisfactory details of a male, both at rest and in flight, observed by H. M. Doubleday at Dawlish presumably in that year, but the record unfortunately omits the date. The only fully authenticated occurrences are that of an adult female which my son and I watched closely on Woodbury Common on 21 May 1961, described in detail in the *Devon Report* for 1961 and recorded in *BB* 55 : 571, and a male seen at Strete near Slapton on 15 August 1967, *BB* 61 : 338.

KESTREL   *Falco tinnunculus*

Resident, breeds

Resident and generally distributed, this falcon was considered by D & M to be outnumbered only by the Sparrow Hawk. It now vastly exceeds the latter and is by far the commonest diurnal bird of prey in Devon. It breeds commonly on the coastal cliffs, is very widespread throughout all types of country and is resident on Exmoor and Dartmoor, though possibly not so common on the highest central areas. On Dartmoor, however, it is known to use some of the tors as breeding sites in addition to the more usual crows' nests in trees.

Of the very few records of actual numbers mentioned in the *Devon Reports*, that of thirteen birds seen in the air together by M. R. Edmonds at Start Point on 22 September 1962 is of particular interest.

In at least thirteen of the years from 1947, one or two pairs have bred on Lundy, where three pairs were believed to have nested in 1957, 1958, and 1959, but none in 1965 and 1966. Breeding on the island was also reported in 1922, 1927, and 1930, and almost certainly occurred in many other years. The species is recorded annually, however, and during most months of the year, with some spring and autumn passage in most years, and a maximum of twenty birds observed on 2 September 1951.

RED GROUSE   *Lagopus lagopus*

Resident, breeds

According to D & M the Red Grouse was introduced on Exmoor in 1820-25 but did not become established. The species was square-bracketed in their list because the only two nineteenth century records for Devon were considered doubtful. It is recorded in *BB* 13 : 86 that the species was reintroduced on Exmoor in 1916, and in 1919 was increasing. The *VCH* contains no further record for Devon, and the first mention of the Red Grouse in the *Devon Reports* is in 1935 when it was stated to be still thinly distributed on Dartmoor some 20 years after its introduction or reintroduction—evidently around 1915.

It is still thinly distributed on Devon Exmoor; a pair with young was recorded at Brendon Two Gates in June 1960, and parties of up to four were reported on Molland Common during 1966 and the early months of 1967.

On Dartmoor it is at present thinly but fairly widely distributed as a breeding bird above 1,300 ft, mainly on the northern half of the moor, but also in very small numbers on the southern part. The *Devon Reports* for 1961-2 each contain records of about forty individuals observed in different localities of Dartmoor. During 1964 parties of up to nine were recorded from thirteen different areas, while in 1967 birds were seen at ten localities in the north and five in the south. In general it may be said that the most likely area for observing Red Grouse is within about a 4 mile radius of Cranmere Pool.

Despite some losses during the severe winter of 1962-3, the numbers do not appear to have been seriously reduced.

BLACK GROUSE   *Lyrurus tetrix*

Resident, doubtful if still breeds

Formerly resident on many moorlands in the county, the Black Grouse has continued to decrease during the present century almost

Page 103: *Common Sandpiper at nest. Formerly a regular breeding bird on Dartmoor and Exmoor, it now occurs chiefly as a common passage migrant on the coast and estuaries*

Page 104: *Bovey Valley National Nature Reserve. A native deciduous woodland on the fringe of Dartmoor; a haunt of Dippers, Grey Wagtails, and warblers*

to the point of extinction, and it is now doubtful whether it still breeds on Dartmoor, though a few may still survive on southern Exmoor.

Palmer and Ballance in *The Birds of Somerset* stated that it was now restricted mainly to north-east Exmoor, but a few might survive on Molland Common in Devon. D'Urban stated in the *VCH* that this species was abundant on most moorlands at the beginning of the nineteenth century, and at its close some still existed on Dartmoor, Haldon, the Blackdown Hills, and the moorlands of north Devon.

During the years 1930-50 it was regularly recorded at several different localities on Dartmoor and on Devon Exmoor. The last record of a Black Grouse on Haldon referred to one killed in March 1944, and the last in east Devon was one seen at Culmstock Beacon in 1946-7. The population in the Halwill area was stated to number about a dozen in 1933, and in 1934 the species was thought to be surviving well under the increasing afforestation in this locality, but there were no subsequent records.

At a 'lek' at Bellever, where seven cocks and two greyhens were recorded in May 1949, the last bird was seen in 1954. Similarly, a few survived until the early 1950s at Fernworthy, but none has been seen since 1953. Six cocks were recorded on south Exmoor in August 1950, and the *Devon Report* records that a few still existed there in 1966.

On northern Dartmoor nine were observed at High Willhays in November 1957, a pair at Okement Hill in October 1959, and one at Gidleigh in 1967.

RED-LEGGED PARTRIDGE   *Alectoris rufa*

Resident, breeds locally

D'Urban wrote in the *VCH* that this species was 'first introduced into Devonshire probably about 1840, and at different times since then some have been turned down in several parts of the county, but although a few examples are met with from time to time, it has never thoroughly established itself'. The position is the same at the present time, for in spite of further introductions the Red-legged Partridge is unable to establish itself in the county.

G

The species has been reported in about 17 of the 40 years since 1928. During the 1930s and 1940s it was fairly regularly seen in the East Budleigh and Lympstone areas of east Devon, but the stock gradually decreased, though occasional birds were still being reported in this area up to 1957. A small colony existed near Hartland Point in and around 1932, and a single bird was killed at Braunton in December 1942. Nesting was reported at Rousdon in 1944 and a pair was seen at Powderham in 1949. In June 1962 three pairs of adults and fifty young were released at South Tawton. The latest record in the *Devon Reports* refers to a single bird which was seen at Dawlish Warren on 21 November 1965, and may possibly have come from the South Tawton stock.

There is no record for Lundy; the species is not mentioned by Ryves as occurring in Cornwall and is scarce and local in Somerset.

PARTRIDGE   *Perdix perdix*

Resident, breeds

D & M's account of this species, including the statement that 600 were shot on Tawstock Court estate in September 1887, suggests that the Partridge was much more plentiful in those days than it is now. Although it now occurs and breeds over much of the county, except for the high ground of Dartmoor and Exmoor, it is definitely more common in the east than the west. Being a favourite of sportsmen rather than of birdwatchers, however, it tends to be dismissed in the *Devon Reports* as : 'a rather poor season' (1946); 'plentiful in all areas' (1958); 'many coveys reported but not very numerous in the west' (1967); and there are practically no records of actual numbers.

It is very rarely recorded on Dartmoor, but a pair is reported as having bred successfully near Postbridge in 1961, and a pair was recorded at Bellever in June 1962. In 1943, B. G. Lampard-Vachell regarded it as being common in the Torrington district. Whilst the numbers are difficult to assess, however, it seems probable that the species has decreased during the past 30-40 years.

According to P. Davis, it has been introduced on several occasions on Lundy during this and the previous centuries, but it has quickly died out. The latest introduction was in August 1966 when five birds were released.

QUAIL   *Coturnix coturnix*

Irregular summer visitor, breeds sporadically

The Quail was definitely much more plentiful in Devon during the nineteenth century than it is at the present time, although the records suggest a slight increase during recent years. Summarising its status in the *VCH*, D'Urban wrote of it as being an irregular summer migrant, some occurring every year, but plentiful in certain years such as 1870 and 1892, when many occurred and numerous pairs bred in different parts of the county, especially in the South Hams. D & M state that the Rev H. G. Heaven is said to have known of thirteen or fourteen nests on Lundy during 1870.

Although the Quail has bred in Devon in at least 4 years since 1900, it can only be regarded as an irregular summer visitor, occurring in some years but not in others. The breeding records which I have been able to trace are as follows: 1904, when three birds of a pair with ten young were shot near Tiverton during early September (D'Urban's MS); and 1934, when Dr F. R. Elliston Wright found a nest with eight eggs in barley stubble on Braunton Great Field, which he recorded in the *Devon Report* for 1934, and illustrated with a photograph in the 1935 *Report*. The species was recorded as nesting at Bradworthy in 1946, when a young bird which had been shot was shown to 'S.C.', who contributed the note. In the following year, 1947, which was a 'quail year', Dr Wright again found and photographed a nest in the Braunton Great Field, the illustration being reproduced in the *Devon Report* for 1947. Whilst breeding was not proved in the 'quail year' of 1964, it is very probable that the species did nest, as birds were present at North Tawton from 18 May until 14 August, and at least two were heard and seen at Zeal Monachorum from 23 June to 14 August, while two were heard at Stokenham during June and July. In D'Urban's time the Great Field at Braunton was a favourite resort of the Quail, and although Dr Wright actually found nests there only in 1934 and 1947, he told me in 1948 that he considered the species had bred there during other recent years.

The records of this species in the *Annual Reports* are very few up to 1942, but it was recorded in the county in 16 of the 24 years from 1943 to 1966 inclusive. It evidently occurred regularly on Lundy during the last century, but it has since decreased there, as elsewhere, and now occurs only occasionally on migration, chiefly during May.

PHEASANT *Phasianus colchicus*

Resident, breeds

An introduced bird, of which the present stock represents the cross-ing of a number of different races, the Pheasant is plentiful on low ground, especially where preserved and artificially reared. From Dartmoor and Exmoor, however, it is absent except for the very occasional bird in cultivated areas. After several attempts at intro-duction on Lundy during this and the previous century, the attempt around 1922 was successful and a population of several pairs has since bred there. Four further birds were released on the island in October 1963.

Dr E. Moore, writing in 1830, described the Pheasant as being abundant in the county at that time.

CRANE *Grus grus*

Vagrant

Until quite recently the inclusion of this extremely rare species rested on two very old records quoted by D & M, the first concerning a single bird which frequented the banks of the River Tamar for several days during the autumn of 1826, and was at last shot at Buckland Monachorum. The second, a sight-record, concerns one which was seen for five or six days in some fields near Start Point during September 1869, and which successfully kept out of gunshot. The record contained in the *Devon Report* for 1944 of three seen on the Otter estuary on 10 September 1944 was thought to refer to birds that had escaped from captivity, although in the light of the most recent records they could well have been genuine wild migrants.

On 16 November 1961 an immature bird of this species was injured by striking overhead wires near Dartington, and was taken to Paignton Zoo where it recovered. W. P. Chubb and H. Collings saw three or four others circling overhead at the time. The record was published by the Rarities Committee in *BB* 55 : 571, where it is noted that parties of four Cranes seen in Cornwall on 15 November and in Lancashire on 11 November may have referred to the same birds.

Of a remarkable movement of some 500 Cranes across the coastal strip of southern England between 29 October and 3 November 1963, seventeen were seen flying over Seaton on 31 October, and were reported by T. J. Wallace, while at West Charleton on the Kingsbridge estuary R. V. Price saw and heard the calls of at least seven flying over in bright moonlight. On the following day seven were seen on fields in the same locality by Rev L. Rotathair and Mrs Milward; these records are detailed in the *Devon Report* for 1963. A paper by D. D. Harber on this extraordinary movement and probable origin of the Cranes in northern Europe was published in *BB* 57 : 502-8.

A further occurrence on 23 September 1966 relates to a single bird seen flying over Slapton Ley by F. R. Smith, L. I. Hamilton, and R. V. Price. What was most probably the same individual was seen earlier in the same day over Woodbury, and later that day over Bowcombe Creek. This record was listed in *BB* 60 : 316 after acceptance by the Rarities Committee.

WATER RAIL   *Rallus aquaticus*

Winter visitor and passage migrant, formerly bred

The Water Rail, according to D & M, was said to breed on Dartmoor, at Slapton, and on Lundy, but their only definite evidence of breeding was a clutch of eggs, in their possession, taken in north Devon. W. Walmesley White reported that two pairs bred on the Otter Marshes until 1918 and thereafter one pair until about 1926, while Dr F. R. Elliston Wright noted it as a breeding species at Braunton.

There is no proof of its having bred in Devon during the past 40 years and in fact there are only about two records of birds being seen during the breeding season, one at St Giles in the Heath on 22 June 1952 and the other at Molland on 17 June 1954.

From August or September, sometimes late July, until early April they occur at Slapton Ley and on the coastal marshes bordering the estuaries, the early arrivals probably being passage birds. In November and December they are usually quite common in the reed beds of the Exe estuary, and they occur regularly in all other suitable marshy localities. The species is rare on high ground, however, and there are apparently no records for Exmoor and few for Dartmoor, though one was seen on the latter at Green Hill, almost in the centre of the southern moor, on 25 December 1951; but P. J. Dare did not record it around Postbridge during the 1950s.

The Water Rail visits Lundy regularly on migration, chiefly on autumn passage, when up to five or six may be seen in a single day during September or October. It occurs less regularly and in smaller numbers in spring, and fairly regularly as a winter visitor in small numbers of up to four or five. There is no substantiated record of breeding on Lundy.

SPOTTED CRAKE   *Porzana porzana*

Irregular visitor

Although many records are listed by D & M for the nineteenth century, the Spotted Crake was subsequently stated in the *VCH* to be much scarcer than formerly. It now occurs chiefly as a rare and irregular autumn passage migrant, of which about twenty examples have been listed in the *Devon Reports* since 1928, all but three being in the years from 1950 to 1967.

The only records for the north refer to one found dead on Braunton Burrows in September 1950 and one heard at Torrington in September 1965. The remainder mostly refer to the Axe and Otter estuaries, three each, and Slapton, where three were ringed in the Septembers of 1962, 1964, and 1965 and about four others seen, including a single wintering bird in 1964-5 and two birds in September 1966. One wintered on the Otter in 1963-4. There is one inland occurrence, at Eggesford in November 1943, and the species was recorded on Lundy in 1887, with a probable occurrence in August 1966. The twenty examples mentioned also include four spring records for the months of March and April.

It still breeds occasionally in Somerset.

LITTLE CRAKE   *Porzana parva*

Vagrant

The first known occurrence of the Little Crake in the British Isles was an example shot near Ashburton in 1809 and described by Montagu as the Little Gallinule, in his Supplement to the *Ornitho-*

logical Dictionary. A further specimen, listed by Dr E. Moore as the Olivaceous Gallinule, was caught by some boys at Devonport on 13 May 1829. D & M relate that three were caught by a dog in a tract of swampy moorland at Frogmore near Ilfracombe, during one winter about the year 1850, and another was picked up at Kingsbridge on 27 July 1855. A further example was shot on Braunton Burrows on 4 February 1876.

There have been few occurrences of this small crake during the present century, and not all of these are adequately authenticated. The first relates to a bird caught in a garden on the edge of Woodbury Common on 21 March 1932, and described in detail by W. Walmesley White who recorded it in the *Devon Report* for that year. He informs me *in litt* that he also recorded one on the Otter estuary on 14 January 1939. The most recent occurrence, and the only one for Lundy, is an adult male that was observed in the walled garden of Millcombe on 12 and 14 September 1952 (p 16 of the *Lundy Report* for 1952).

CORNCRAKE   *Crex crex*
Passage migrant, formerly bred

D & M, who described the Corncrake as a common bird, stated that it was plentiful on the hill farms in north Devon, and though scarcer as a breeding bird in the south, it occurred commonly during the autumn near the south coast; for instance, eighty-four were shot at Malborough on 23 September 1861. They were not infrequently encountered on Dartmoor during the autumn at such places as Raybarrow Mire, and they occasionally wintered in the county.

In 1906 D'Urban noted that it still frequently bred in the north, and Loyd recorded breeding at Branscombe, Seaton, and Beer prior to 1929, but noted that it was annually becoming scarcer. The *Devon Reports* state that four pairs bred in the Exmouth area, one at Bradiford and another at Holsworthy in 1932, in which year F. R. Elliston Wright counted at least thirty newly arrived birds near Saunton. A pair were reported to have bred near Torquay in 1933, a brood was seen at Dartmouth in 1937, and three pairs were believed to have nested there in 1938. This appears to be the last breeding record for the county, although the Corncrake continued to be seen annually, mainly in spring and autumn, and it seems probable that it continued for several years to breed at Braunton if not elsewhere.

Since 1950 there have been about sixty-five records from all parts of the mainland during the months from February to November, including five for April, ten May, nine June, fifteen August, and eighteen September, of which some of the June birds could possibly have bred.

A regular spring and autumn passage migrant in very small numbers on Lundy, it is said to have bred there regularly during the last century and was recorded as breeding in 1928 and possibly 1935.

MOORHEN   *Gallinula chloropus*
Resident, breeds

An abundant breeding bird in Devon, the Moorhen is widely distributed on still and slow-flowing water, occurring wherever there is a small pond, river, stream, or dyke. Frequenting freshwater, it is usually absent from the lower reaches of the estuaries but plentiful in tidal water higher up, and is abundant on the estuarine marshes at Exminster, Braunton, and elsewhere. A party of forty-three, however, was recorded on the Torridge at Bideford Bridge on 21 February 1955. It occurs abundantly in the lowland reaches of the rivers and at Slapton and Beesands Leys.

On Dartmoor, although it is occasionally seen at Fernworthy and Burrator reservoirs, it is absent from the fast-flowing rivers and was not recorded by P. J. Dare in the Postbridge district during the 1950s. Similarly, it has been recorded at Wistlandpound on the fringe of Exmoor, but is normally absent from high ground.

Resident and for the most part sedentary, the Moorhen shows practically no signs of migratory movement. That some movements do occur, however, is evidenced by the individuals that occasionally visit Lundy, where about nine singles have been recorded in various months since 1950. Davis records that unsuccessful attempts to introduce it on Lundy were made in 1937 and 1938.

Although there is no evidence of any great decrease since the last century, the continuing loss of habitat through land drainage may well be reducing its numbers.

COOT *Fulica atra*

Resident and winter visitor, breeds

As a breeding species, the Coot is resident on most of the larger sheets of freshwater, but does not nest on the Dartmoor reservoirs, though it visits them in small numbers during the winter. Breeding by two or three pairs on each water occurs regularly or has been recorded in recent years on Tamar Lake and Horsey Ponds in the north, and Stover, Creedy Pond, Countess Wear, Kitley Pond, and Beesands Ley in the south; and up to about twenty pairs have bred on the old canal between Tiverton and Holcombe Rogus. The main concentration of breeding birds, however, is at Slapton Ley where the number has been variously recorded as from twenty-five to fifty pairs.

During the winter the resident birds are greatly outnumbered by visitors whose numbers fluctuate according to the severity of the weather. Severe conditions bring flocks on to the estuaries, too, where in mild winters few, if any, are seen. In addition to fluctuations due to the weather, there also seems to be a tendency for this species to change from one locality to another over a period of years. The changes on the Exe estuary in particular are very marked, for according to D & M the Coot was abundant on this estuary in the mid-nineteenth century, but had become scarce by the close of the century. Subsequent records show that it again wintered in large numbers of up to 2,000 birds, subject to the usual fluctuations, during the 1920s and 1930s, after which the numbers declined. Thus, there were about 2,000 in December 1933; 1,000 in 1935; 800 in December 1938; 300 during December 1941; 800 in December 1946; and 1,000 during the very cold spell of February 1947. After that the numbers dropped right down to under twenty for several years, then varied from less than a dozen to 300 during the 1950s. During the present decade the number wintering on the Exe has varied from about 100 to 400.

Slapton Ley, which has always been an important wintering as well as breeding area, nowadays holds from about 200 to 500 birds in normal winters, the numbers increasing to as many as 2,000 in severe winters like 1947 and 3,000 during the first three months of the arctic winter of 1963.

The other waters on which fair numbers nowadays winter regularly are Kingsbridge estuary, up to 600; Beesands Ley, up to 400; Tamar Lake, up to 200; and Wistlandpound reservoir, up to about 100; while flocks of 200-300 have occasionally been recorded on the Taw estuary and Hennock reservoir. Small parties of under twenty-five birds occur irregularly at Burrator and Fernworthy reservoirs on Dartmoor, while the lakes at Stover, Kitley, and Shobrooke hold varying numbers of up to about fifty birds. Small flocks also occur irregularly on the other estuaries and reservoirs throughout the county, and in severe winters like 1963 they are observed at times on the sea at various points along the south coast.

Until 1927 the Coot had not been recorded on Lundy, but since then there have been about nine occurrences, all of single birds and all between the months of October and March, except for one on 13 August 1966.

GREAT BUSTARD   *Otis tarda*

Vagrant

The history of the Great Bustard in Devon is related in some detail by D & M, whose account includes the occurrence of a flock of seven or eight birds at Croyde and Braunton on 31 December 1871. Two of these birds were shot and went into the collections of Cecil Smith and Rev M. A. Mathew. The Devon occurrences appear to have been overlooked by the editors of *The Handbook* in their records of the immigration of this species in the winter of 1870-71. The previous record for the county refers to a female killed at Bratton Clovelly on 31 December 1851. The only other known occurrences are four mentioned by Montagu as having been obtained near Plymouth during the years from 1798 to 1804. D & M themselves described the species as 'a casual visitor, of very rare occurrence in winter'.

In his *Braunton: A Few Nature Notes* Dr F. R. Elliston Wright refers briefly to a record of a Great Bustard in that district in the year 1873, but no details are given and there is no mention of this record in D & M's authoritative work. This species has not been recorded in Devon during the present century.

LITTLE BUSTARD *Otis tetrax*
Vagrant

D & M stated that at least ten Little Bustards had been recorded in Devon, but this was amended in the *VCH* in which D'Urban gave the number as at least a dozen, having evidently included a party of three which were seen on Braunton Burrows in November 1893. He adds that all were recorded during the winter; five occurred in the South Hams and the rest in north Devon.

Although there is no mention of this species in the whole series of the *Devon Reports* (other than in connection with a vague Somerset record in 1945), it has been recorded once during the present century. *BB* 6 : 225 relates that one was shot at Braunton on 11 January 1912 and taken to James Rowe, the Barnstaple taxidermist. None of the Devon examples has been critically examined and assigned to either the eastern or western races of this species.

OYSTERCATCHER *Haematopus ostralegus*
Resident and winter visitor, breeds

The Oystercatcher, which has increased enormously in numbers since the close of last century, was considered by D & M to be principally a passage migrant, though also a winter visitor to the Taw estuary and a breeding bird on Lundy. D'Urban subsequently reported an increase and noted that it bred at one point on the south coast.

P. J. Dare in *Fishery Investigations*, series 2, vol 25, No 5 gives the wintering population for Devon for 1963-4 and 1964-5 as 2,500-5,000 birds, of which 1,500-2,500 occur on the Exe estuary, 500-1,000 on the Taw and Torridge estuary, 200-300 on the Teign estuary, and flocks of up to 100 regularly on the Kingsbridge, Plym, Tamar-Tavy estuaries, Wembury and elsewhere. The breeding population is stated as 25-50 pairs, of which 10-20 bred on Lundy, 10-20 on the north coast, mainly in the Hartland Point area, and 5-10 pairs were scattered along the south coast.

Breeding dates listed in the *Devon Reports* include : Scabbacombe 1950, Taw estuary 1957, Braunton Burrows 1960, Start Point 1961, Budleigh Salterton 1962, Bull Point 1963, and regular dates for Morte Point and Lee. Loyd recorded breeding at Beer Head in 1927.

The peak numbers occur between August and October on the Exe estuary where a maximum of 3,800 was recorded in September 1966, while on the Taw estuary 1,600 were recorded in October 1964. Many non-breeding birds remain on the estuaries throughout the summer, with a maximum of 1,500 on the Exe in June 1966. They are infrequent inland.

Birds are present on Lundy throughout the year but usually in smaller numbers during the winter months from October to February.

SOCIABLE PLOVER   *Vanellus gregarius*
Vagrant

The only record of this Asiatic species is of a single bird which frequented Braunton Marsh in the autumn of 1963, when it was seen by a number of observers from 23 September until 12 November. It associated with Golden Plover and Lapwings in the fields at Horsey. A long-legged, sandy coloured bird with a noticeable white eyestripe, it was inconspicuous on the ground but in flight was immediately transformed into a black and white bird with conspicuous white secondaries and black primaries, giving a distinctive appearance to its slender pointed wings which reminded me at the time of a Sabine's Gull. This record, which constitutes the eighth for Britain, is reported in *BB* 57 : 267.

LAPWING   *Vanellus vanellus*
Resident, winter visitor and passage migrant, breeds

Although an abundant winter visitor and still widespread as a breeding species, the breeding population has decreased during the past 20-30 years and was further severely reduced by the exceptionally hard winter of 1962-3. In the early 1900s D'Urban considered it to be increasing as a resident, and it was stated by Bruce Cummings in 1906 to be nesting in hundreds on Braunton Burrows, where it still bred commonly in the 1930s; but only twenty pairs were reported in 1960. The numbers have fluctuated, however, with decreases due to hard winters being followed by recoveries.

Dare and Hamilton describe it as widely dispersed over bogs and wet commons on Dartmoor, with fifteen to twenty-five nesting pairs in the Postbridge district up to 1962, none seen in 1963, and a gradual recovery to five to ten pairs in 1967. Breeding pairs are thinly spread throughout the county on farmland, hill pastures, and grazing marshes. At North Tawton, for instance, there were up to twenty pairs and near Axmouth about ten pairs in 1966.

An abundant winter visitor, it occurs in large flocks on farmland, moving to the estuaries and coastal marshes in cold weather. Hard weather movements often involve flocks of thousands; for example, 15,000 were concentrated on the Axe estuary in mid-January 1964 after heavy snow farther east. Similarly, in 1958 a large influx took place in late December, when some 15,000 were reported on the Exe and Clyst Marshes, in addition to large flocks elsewhere. Winter flocks of 1,000-3,000 occur regularly on estuarine marshes.

Breeding, previously irregular on Lundy, has been annual since 1927, varying from three pairs in 1930, ten in 1939, and two in 1947, to eighteen in 1961 and 1966. Birds are present during most months, with peaks of up to about 140 on spring and autumn passage.

RINGED PLOVER *Charadrius hiaticula*

Resident, passage migrant and winter visitor; breeds

Very small numbers of not more than two or three pairs still breed at three or four places, including the estuaries of the Exe and the Taw and Torridge; six pairs bred on the Taw in 1960 and the number may now be only slightly less. According to D & M it bred at many localities along the south coast and commonly on Braunton Burrows, while Loyd mentions breeding on the Axe and at Beer Head during the 1910s; the decrease is due to disturbance.

This plover occurs commonly on the estuaries at practically all times of the year, the greatest number being present at the height of the autumn passage during late August and early September when 300-800 are reported on the Exe, up to about 150 on the Taw, and flocks of up to about thirty at the other south coast estuaries and Wembury. Smaller numbers are present throughout the winter, including about fifty on the Exe, but there appears to be no marked spring passage.

Maximum counts on the Exe during recent years include 500 in late August 1952, 400 on 4 September 1954, about 800 on 28 August 1960, 526 in August 1964, and 600-700 on 19 August 1967. It is rarely recorded inland, but one was seen at Burrator on 27 August 1955 and a bird found at Cheriton Fitzpaine on 9 January 1935 was of the Arctic race, *C.h.tundrae* (*BB* 31:356), as were five out of six birds trapped on the Exe in September 1961 (*Devon Birds* 14:37).

Up to three or four occur regularly in autumn and irregularly in spring on Lundy where it was thought to have bred in 1942 (*BB* 38:187).

LITTLE RINGED PLOVER   *Charadrius dubius*
Vagrant

This species, which first bred in England in 1938 and has since become established as a breeding bird in much of the eastern half of England, is still no more than a vagrant to Devon where it has been reliably recorded on five occasions. The first known occurrence in the county refers to a bird which was observed on a fresh marsh at Powderham on the Exe estuary from 3 to 12 April 1949. It was identified by F. R. Smith and subsequently seen by R. G. Adams, and is fully documented in *BB* 42:252. The second record relates to another seen on the Exe estuary by R. G. Adams on 20 May 1956. Its appearance coincided with a considerable passage of Dunlin, Sanderling, and Ringed Plover, and the bird is carefully described in the *Devon Report* for 1956. The third and fourth occurrences, also on the south coast, refer to a single bird seen on the Axe estuary by R. T. Cottrill on 22 April 1964, and a party of four was watched both at rest and in flight on the River Tavy by P. Harrison on 15 August of the same year. Both records are included in the *Devon Report* for 1964, and both observers remark on the characteristic flight call, in addition to other details. Lastly, one was observed at Chelson Meadow on 11 April 1967 by R. Burridge.

D & M knew of no Devon occurrence of this charming and steadily increasing species, nor has it so far been recorded either on Lundy or the north Devon coast.

KENTISH PLOVER   *Charadrius alexandrinus*
Scarce passage migrant

D & M knew of only two occurrences, involving three individuals, of this beautiful little plover: a pair in May 1861 and an immature

bird in the autumn of 1875, all from Plymouth. Apart from an in-
definite record in 1945, there is no further occurrence until 1948,
when that most reliable naturalist, R. G. Adams, identified one at
Exton on the Exe estuary, a full account of which is given in *BB*
41 : 249. The fact that a female wintered in exactly the same locality
for four successive winters strongly supports the view that it was
the same individual. This bird, which was first seen on 2 January
1948, remained until 13 March. It returned on 17 October 1948 and
was present throughout the winter until 20 March 1949, returning
again on 27 October to the same stretch of foreshore, where it stayed
until 26 March 1950. It came back for the fourth consecutive winter
on 7 October and was regularly observed up to 18 February 1951,
when it was seen for the last time (*BB* 42 : 94 and 43 : 96).

In addition to the wintering bird, twenty others have been reported in the county during the past twenty years—all singles except
for three occurrences each of two. Of these twenty birds, all of
which are well documented in the *Devon Reports*, seven occurred
during April and May, and thirteen during August to October. One
record, involving two birds, refers to the Taw estuary (10 September
1950), one to the Otter estuary (2 October 1949), one to the Kingsbridge estuary (17 September 1960), and the remainder to the Exe
estuary, where the species has been observed in 13 out of the past 20
years. There is no record of the Kentish Plover ever having occurred
on Lundy.

KILLDEER PLOVER   *Charadrius vociferus*

Vagrant

Charles Dixon in *Bird Life in a Southern County* states that on
7 September 1898 he flushed a Killdeer Plover from Paignton sands.
'There could be no possible doubt about the species,' he wrote, 'for
it rose in a very leisurely way from our very feet, the chestnut-buff
of the rump and tail coverts catching the eye at once.' This was
published after D & M's *The Birds of Devon*, which contains no
record of this American species. Later, in the *VCH* D'Urban noted
this record in brackets.

The only really satisfactory record of this species in Devon is that
contained in *BB* 32 : 372 in which C. R. Stonor relates that he pur-

chased a Killdeer Plover in Smithfield Market on 6 January 1939, and on getting in touch with the dealer who sent it up and through him with the farmer who shot it was informed that it had been seen about the farm for about a month before it was shot. The bird, a male, is preserved in the British Museum, and was obtained at Meeth near Hatherleigh in the first week of January 1939.

The *Devon Report* for 1943 contains a record, reprinted from the *Ibis*, of one said to have been seen between Lympstone and Woodbury Road on the Exe estuary on 1 April 1943 by Willoughby P. Lowe, who stated he was familiar with the species in the United States.

GREY PLOVER *Pluvialis squatarola*
Winter visitor

Although now a plentiful winter visitor and most probably a passage migrant, the Grey Plover was scarce in D'Urban's time, when it occurred only singly, in pairs, or rarely in small flocks. Bruce Cummings, however, noted that it was common on the Taw estuary in the autumn of 1906. In the early 1930s it was still scarce on the Exe estuary where even single birds were recorded, and R. G. Adams noted in 1933 that a few usually occur in winter.

In 1936 the maximum on the Exe had reached nearly fifty, in February 1943 it was about sixty, and in January 1948 about 100. During the 1950s it ranged from seventy-five to a hundred, and in most years since has been 100-150, with a peak of 160 on 21 February 1965. In a number of recent years birds have been present in every month, with up to sixteen non-breeders remaining throughout the summer. During May, or in August when the first returning birds arrive, up to a dozen or more may be seen in their strikingly beautiful breeding plumage.

Flocks of up to thirty, though usually half that, occur annually on the Taw and Torridge estuary, and rather larger numbers are seen on the shingle ridge at Slapton, where fifty were reported on 5 February 1956. Small parties occur on all the other estuaries and at Wembury Point, but there are no records at all of occurrences inland.

The Grey Plover visits Lundy irregularly : about six singles and a party of seven on 12 August 1959 have been reported there since 1947.

GOLDEN PLOVER *Pluvialis apricaria*
Resident and winter visitor, breeds

An abundant winter visitor and passage migrant, the Golden Plover also breeds in very small numbers on Dartmoor. D & M were uncertain about its status on Dartmoor, and long suspected proof of breeding was not obtained until June 1950, when L. H. Hurrell and E. D. Williams closely watched a pair with three downy young in the Cranmere area, where eleven adults were present. Breeding behaviour had been observed in several previous years, and was again seen subsequently, in addition to which, broken egg shells were found in 1960, two young were seen in 1965, and in 1967 four pairs were present in the breeding area. Twenty-four birds at Powder Mills on 24 August 1957 may well have been local stock.

As a winter visitor, the Golden Plover occurs commonly from mid-September to late April in flocks of up to several hundred on Dartmoor, Exmoor, the Blackdown Hills, and on rough grazing land in much of west and north Devon. In hard weather they usually move to lower ground and large flocks congregate on the estuaries and coastal marshes.

Flocks of up to 2,000 occur regularly on the Tamar near Weir Quay, where 5,000 were recorded in November 1958 and 4,000 in February 1964. Since 1940 flocks of about 2,000 have been observed on Braunton Marsh, near Torrington, on south-east Dartmoor, at Tavy Cleave, on Kingsbridge estuary, at Cox Tor, Exminster Marshes, and at Smeathorpe on the Blackdown Hills.

A fairly large spring passage occurs in late March and early April, when numbers of the Northern race (*P.a.altifrons*) in breeding plumage are identified in many localities, eg about 350 of a flock of 1,000 at Smeathorpe on 1 April 1961 were considered to be of this race. One ringed in north-west Iceland in July 1935 was recovered in Devon in November 1935, and another ringed in Iceland in June 1957 was recovered at Buckland Monachorum in March 1958.

It is a common passage migrant on Lundy where it has been observed in all months. The most recorded there appears to be 100 in January 1955.

H

DOTTEREL   *Eudromias morinellus*

Passage migrant

It is somewhat surprising that D & M were able to list only about a dozen or so occurrences of this species during the whole of the last century. Summarising its status in the *VCH* D'Urban wrote: 'The Dotterel rarely visits Devonshire, and generally only on passage in spring and autumn, but a few have occurred in the winter months'. Regular observation on Lundy has shown that this species visits the island fairly frequently, though not annually, having been reported there in 11 of the 20 years from 1947 to 1966. While most of the birds have been recorded on autumn passage in late August and early September, there are also six spring records for the island during this period, divided equally between April and May. In most years only one or two birds are seen, but a party of eight was recorded on 14 April 1949, three from 7 to 12 September 1960, and up to five from 1 to 5 September 1961. The species was particularly well recorded during 1966, with one on 27 and two on 28 April, five from 7 to 10 May, and one which was ringed on 7 August and remained on the island until 22 August. In addition to these records, W. B. Alexander reported in *BB* 36 : 140 having seen a single bird on Lundy on 9 and 10 September 1942 and mentioned that F. W. Gade had two spring records of Dotterel on Lundy.

In comparison with the Lundy records, those for the mainland of Devon during the present century are surprisingly few. They comprise a single bird on Budleigh Salterton golf course for about two weeks from 17 September 1920, a small flock near Little Torrington on 22 March 1932, single birds on Ugborough Beacon in August 1944, at Tamar Lake on 27 July 1959, Dawlish Warren on 10 September 1961, and Hameldown, Dartmoor on 18 September of the same year, and, lastly, two on the Teignmouth golf course on 5 September 1965. In addition, J. Coleman-Cooke informs me that he saw two Dotterel close to the White House on Braunton Burrows during one spring about 1958.

It is apparent from the Lundy records that this very confiding plover has always visited the island on passage migration but due to the absence of observers until 1947 has passed unnoticed except for the occasional record.

TURNSTONE   *Arenaria interpres*

Passage migrant, winter visitor, and non-breeding summer visitor

Whereas D & M regarded the Turnstone only as a passage migrant, it is nowadays also a common winter visitor, and in smaller numbers a non-breeding summer visitor. It has increased since 1900, particularly so during the past decade, and now occurs in hundreds on the Exe estuary where it was formerly counted in tens. The two main strongholds are the Exe estuary and Wembury Point, at both of which it is present throughout the year—in its greatest numbers during May and September when passage birds are present, and in its smallest during June when only non-breeders are left.

Some of the maximum counts recorded during recent years include 125 at Northam Burrows in September 1958; 300 at Wembury in November 1961; 140 on the Exe in April 1962 and 400 at Wembury in October of that year; 320 on the Exe in August, over 500 at Wembury on 12 September, and 170 at Prawle Point on 9 September, all in 1964; and 450 on the Exe in May, 100 on the Avon in December, and 500 at Wembury in November 1966. Smaller numbers occur at many other points along both the north and south coasts, but it is most unusual inland, where the few occurrences probably relate to storm-driven birds.

Varying numbers of up to about twenty occur on Lundy, possibly chiefly on spring and autumn passage, but in practically all months. A bird ringed at Wembury in January 1951 was found breeding on Ellesmere Island in June 1955.

LONG-BILLED DOWITCHER   *Limnodromus scolopaceus*

Vagrant

Under the name of the Red-breasted Snipe, four Devon occurrences are enumerated by D & M. The first relates to a male shot on the coast of Devon in October of some year about 1801. Colonel Montagu, who first made this species known as a British bird, described it in his *Ornithological Dictionary* as the Brown Snipe. This particular specimen, which is now in the British Museum, has recently been critically examined and proved to be a Long-billed Dowitcher, and is the only definite record of *L.scolopaceus* for the county. Of the three

other records of dowitchers cited by D & M, one refers to a bird shot
in the parish of South Huish in the winter of 1855; another undated
example was said to have been killed at Hatherleigh; while a third,
which was in the collection of a Mr Drew of Devonport, bears
neither date nor locality, but is assumed to have been taken in Devon
in about 1837. This and the South Huish example are listed by I. C.
T. Nisbet in *BB* 54 : 340-357 as unidentified between Long- and Short-
billed Dowitchers. As there is no subsequent record of either species
in Devon, the Short-billed (*L.griseus*) cannot be included in the
county list, although one of the indeterminate records could well
have referred to the latter species.

Both are North American waders which occur in Britain as very
rare autumn vagrants.

SNIPE   *Gallinago gallinago*
Resident and winter visitor, breeds

D & M remarked on the decrease in the number of Snipe during the
second half of last century, but stated that it still bred in various
parts of the county. At the present time it breeds principally on the
Dartmoor bogs, but probably also on Exmoor, where nesting was
recorded at Challacombe in 1932, though there are no recent records.
Dare and Hamilton give the present breeding population for the Postbridge district of Dartmoor as fluctuating to a maximum of fifteen
to twenty pairs, and they record a decline in numbers during the
past thirty years.

It almost certainly still breeds on Braunton Marsh, where nesting
was proved during the 1930s and breeding behaviour has been recorded in the 1960s. During the 1950s it was nesting regularly on
Woodbury Common. Although there are nowadays few records of
nesting in other localities, it must certainly breed in some of the
upland marshes of west Devon, at one of which, Thornhill Head,
drumming was reported in May 1961.

The Snipe occurs most plentifully as a winter visitor, being common and sometimes abundant on the estuarine marshes, and frequent
in bogs and marshes throughout the county, but moving to lower
ground in hard weather. Numbers of up to 100 are commonly reported in many areas between September and March, but in more

favoured localities such as the Axe, Exe, and Clyst Marshes many more occur. In the winter of 1959-60, for instance, 300 were regularly roosting in the reed beds at Topsham, and in both winter periods of 1962, 300 were recorded on the Axe Marshes. The most recorded in recent years was 400 on the Tamar on 27 December 1955.

Varying numbers visit Lundy during spring, autumn, and winter, while breeding was suspected in 1930 and 1935. The maximum counts since 1947 were forty-five in January 1955 and a peak of forty-eight on 23 March 1962.

GREAT SNIPE *Gallinago media*
Vagrant

The general decrease of the Great Snipe is reflected in its very infrequent occurrence in Devon during the present century, compared with former records. In fact, the satisfactorily documented occurrences since D & M's day are extremely few. In the *VCH* D'Urban wrote of it as being 'a casual visitor occasionally met with on the moors' and stated that in October 1868 no less than seven occurred in the county, while in their more detailed book D & M cite fifteen dated records, mostly of birds shot between the years from 1846 to 1886.

The *Devon Reports* contain five records, one of which refers to west Somerset and is therefore irrelevant. Two of the other records are completely lacking in evidence of correct identification, so are of little value. These two records simply state that one occurred at Hatherleigh on 9 March 1931 and that another was seen at North Molton in December 1933. The next record, which was square bracketed as being considered doubtful, is in my opinion more satisfactory than the two previous ones. It concerns two birds observed by Mrs F. E. Carter at Tamar Lake on 22 August 1949, the description of which leaves little doubt of their being correctly identified. The most recent record relates to a bird of this species seen by S. D. Gibbard beside the Otter estuary on 23 February 1956 and accurately described in the *Devon Report* for 1956.

Although D & M mentioned that one was killed on Lundy, neither date nor details are given and the record was rightly square bracketed in Peter Davis's *List of the Birds of Lundy*, for which place there is no satisfactory record.

JACK SNIPE   *Lymnocryptes minima*

Winter visitor

This very small snipe is a regular winter visitor, occurring singly or in parties of three or four, and preferring certain localities year after year. The species is widespread and occurs at bogs and marshes in all parts of the county, both at sea level and on the highest parts of Dartmoor, and equally in the north and south. Because of its habit of sitting very close and often rising almost from underfoot, this unobtrusive bird must frequently pass unnoticed and doubtless it occurs far more frequently than the records suggest.

From about ten to twenty-five individuals are recorded on the mainland annually, mostly from very scattered localities, but almost a dozen and frequently six or more have been flushed from a small marsh at Dawlish Warren, where eleven were reported in March 1961, ten in February 1966, and seven in December 1960. Seven were recorded on the Axe on 22 December 1962. Four or five at a time have been seen at Chelson Meadow, Plymouth; on the Braunton and Clyst Marshes; and on the Tamar: while thirteen were flushed from a small marsh at Dartmouth in January 1936.

Most of the occurrences are in the months of December to February, but they range from September to mid-April. The species visits Lundy almost annually in small numbers, with occasional records of up to four birds, but usually no more than about six during the course of a year. No change of status is apparent since the last century.

WOODCOCK   *Scolopax rusticola*

Winter visitor, has bred

A regular winter visitor which has occasionally bred, the crepuscular Woodcock is common in some years but scarce in others, being generally more frequent in severe winters, when it occurs in most woodlands, often on bracken-covered hillsides, and particularly in the deep, wooded river valleys.

D & M knew of only a few breeding records: Whitstone Wood 1853; presumed nesting in France Wood, Stokeley 1891; and three nests at Tetcott 1892. W. B. Alexander in *The Woodcock in the*

British Isles stated that it nested near Budleigh Salterton in 1929, near Kingsbridge in 1932, near Honiton in 1933, and was believed to breed annually on the Blackdown Hills. Loyd stated that it bred regularly near Branscombe. G. H. Gush informed me that it nested in Ashclyst Forest in 1934-5. The *Devon Reports* record that young were seen at Lapford in 1932, 'roding' was observed at Mary Tavy in February 1934, and the 1964 *Report* mentions that a nest with two eggs was found at Meavy 'about fifty years ago'. Breeding was suspected at Monkleigh in 1955. Prey at a Buzzard's nest near Moretonhampstead in 1966 was reported to include a young Woodcock—*Devon Trust* 11 : 434.

As a winter visitor it occurs from mid-October until March or early April. Alexander gave the winter population of six areas of woodland near Kingsbridge as 100-120 birds and up to eighty in 15 square miles around Exmouth. In the wooded Torridge valley near Merton, G. H. Gush reported ninety in March 1950 and twenty were recorded at Arlington in December 1951. Many were reported in all areas during the hard winter of 1962-3, including twenty at Baggy Point in January.

On Dartmoor it occurs annually, chiefly in the conifer plantations and the fringe woodlands, but it has been seen at Piles Copse and Wistman's Wood.

Formerly more abundant on Lundy, it now occurs annually as a passage migrant and winter visitor between October and March, but in very small numbers not exceeding four in one day.

CURLEW *Numenius arquata*

Resident and winter visitor, breeds

The Curlew increased very considerably as a breeding species during the present century, and by the early 1960s was nesting commonly throughout the county, not only on the moors, downlands, rough pastures and upland marshes, but also in some lowland areas and on cultivated farmland. Nesting on Braunton Burrows was reported in 1958, and by 1960 had increased to seven pairs. There are breeding records from the Blackdown Hills to Tamar Lake and from Ilfracombe to Dartmouth.

Heavy mortality was caused by the winter of 1962-3, from which the Curlew has not yet fully recovered.

Summer visitors to their inland breeding grounds, where they are present from early March to about mid-August, they congregate in autumn and winter on the estuaries and adjacent marshes, where some non-breeding birds are present throughout the summer. Peak numbers occur in August and September: an exceptional number of 4,000 was reported in August 1950 on the Exe estuary where counts of 3,000 were recorded in September 1947 and August 1952, and in recent years there were around 2,000, of which about half winter. Autumn flocks of 1,000 have occasionally been recorded on the Axe, Taw, and Kingsbridge estuaries, but the usual winter population of most estuaries is 100-400.

Curlews occur on Lundy in all months but chiefly in spring and autumn. A pair has nested irregularly since 1940 and very occasionally before. Breeding was recorded in about thirteen of the years 1947-66, with possibly two pairs in 1959 and one successfully in 1966. A passage of 120 birds on 11/12 July 1962 appears to be the most recorded.

WHIMBREL   *Numenius phaeopus*
Passage migrant

A common passage migrant, the Whimbrel occurs regularly on both spring and autumn passage, the birds travelling overland across the county. The spring passage lasts from about mid-April to late May, and the autumn passage from late July until the end of September; but because of occasional summering and wintering birds and late migrants the species has been recorded in every month.

The numbers vary from one year to another, and counts on the Exe estuary gave maximum numbers of 300 on 27 April 1946, 400-500 on 29 April 1951, 200 at Countess Wear on 6 May 1951, and 100 on the estuary on 21 April 1952; but they are usually well below 100, although many birds must pass through during the course of a season. Flocks of up to about sixty are recorded on the Taw estuary and rather smaller numbers at Slapton and the other south coast estuaries, at all of which the species is regularly reported.

Passage overland is proved by the many records of flocks moving northwards up the Exe, passing over North Tawton and entering the

Taw estuary from the south in the spring, and the reverse in autumn. There are also many records of small parties observed on different parts of Dartmoor and occasional birds at Tamar Lake and Wistland-pound.

The Whimbrel occurs regularly on Lundy on both passages, with maximum numbers of around fifty and regular flocks of ten to twenty. A bird considered on wing length to be of the Icelandic race, *N.p.islandicus*, was trapped on the Exe estuary by D. B. Cabot on 26 August 1961, *Devon Birds* 14:39. The present status indicates no change since last century.

BLACK-TAILED GODWIT *Limosa limosa*
Winter visitor and passage migrant

A spectacular addition to the present day avifauna, the Black-tailed Godwit was a rare bird at the end of the nineteenth century, and one whose every occurrence in the county merited a note in *British Birds* from then until about 1930. In September 1933 H. G. Hurrell recorded fourteen on the Tavy estuary, and R. G. Adams forty to fifty on the Exe in July 1935. By September 1944 the number on the Exe had reached 104 and the species was being recorded regularly on other estuaries. In 1946 R. G. Adams recorded having seen them in every month, and in April 1949 R. M. Curber counted sixty on the Tavy, and the species was already established as a winter visitor and passage migrant.

In late October 1955 there were 300 present on the Tavy and the number continued to increase in the Plymouth area, where the flocks move between the Plym, Tavy, and Tamar. On the Plym estuary 464 were recorded in January 1962. Meanwhile, the number on the Exe had soared to 360 in December 1960, and to 527 on the Plym in November 1963.

In 1966 a total of about 600 were wintering in the county, with some on practically every estuary, and birds occurring during every month. In December 1966, 430 were counted on the Exe and 200 on the Plym, with smaller numbers elsewhere. Inland occurrences are unusual, but the species has been seen several times at Tamar Lake during recent years. It is also infrequent on Lundy where from one to three birds have been observed on about eight occasions since 1938.

BAR-TAILED GODWIT *Limosa lapponica*

Passage migrant and winter visitor

The status of this species in Devon has changed dramatically since D & M's time, as it now occurs not only as a passage migrant but also a common winter visitor and non-breeding summer visitor. On the Exe estuary, where the vast majority occur, they are nowadays present in every month of the year. The wintering flock has gradually increased from about twenty in the early 1930s to 103 in February 1939, 200 in December 1946, 300 in February 1950, 400 in January 1962, over 800 in January 1966, and 700 in January 1967. No doubt the increase has been assisted by the protection afforded there and the general decline in shore shooting.

From 50 to 100 non-breeding birds are recorded in June, the peak so far being 103 in June 1964. The number recorded on spring passage varies, but in April and May 1957 it was 233 of which 114 were in their red breeding plumage, while in April 1962 the total was over 100. Smaller numbers from a few to about thirty, sometimes more, occur in the autumn, and occasionally winter, on most of the other estuaries and at such points on the coast as Hope's Nose and Wembury, while in the north the species is fairly regular on the Taw estuary where they numbered eighty in the severe winter of 1963. This bird is very rarely recorded inland, but five were seen at Tamar Lake in August 1962. Likewise, it is infrequent on Lundy, where ones and twos on spring or autumn passage have been reported in about six of the years since 1947.

GREEN SANDPIPER *Tringa ochropus*

Passage migrant and winter visitor

A regular passage migrant, more common in the autumn than the spring, the Green Sandpiper visits the creeks and fresh marshes bordering most of the estuaries, and occurs at many inland localities, including Dartmoor—where it visits the reservoirs and has been recorded on many of the moorland rivers and pools. In the north it occurs annually on the marshes at Bishops Tawton and Braunton and is fairly frequently reported from Tamar Lake and the reservoirs at Melbury and Wistlandpound.

In many years during the past two decades, thirty to fifty birds have been recorded, usually singly, but quite often in small parties of three or four and sometimes up to five or six, while O. D. Hunt saw a flock of twelve at Newton Ferrers on 10 August 1949. Parties of five have quite often been recorded on the Axe and Exe estuaries, where the species occurs regularly, and six have been seen together on the Erme.

Most years a few individuals remain throughout the winter, particularly on the Taw and Clyst Marshes, at each of which four or five have been recorded in a number of recent years, while occasional wintering birds are reported from time to time at most of the sheltered creeks.

Whilst more birds occur during August than in any other month, the autumn passage covers the months of July to September, followed by wintering birds in October to March or April, a small spring passage in April and May, and the very occasional non-breeding bird in June. The Green Sandpiper occurs as a regular passage migrant on Lundy annually in the autumn, when up to eight or nine singles may be recorded during July to September, and in some years during the spring, when a few singles may pass through during April and May.

Because of the occasional summer records, D & M were firmly though erroneously convinced that this species sometimes nested in Devon, but they were never able to establish it as a fact. The status of this very fascinating sandpiper does not appear to have changed since their time.

WOOD SANDPIPER   *Tringa glareola*
Passage migrant

In view of the regularity of its visits during the past twenty years it is surprising that so few examples of this graceful sandpiper were recorded during the nineteenth and the first half of the present centuries. D'Urban in the *VCH* described it as being very scarce in Devon, and gave details of only six birds, while it is not even mentioned in Loyd's *BSED*, published in 1929, or in his list of the birds of Lundy.

Of more than ninety birds recorded during the present century, all but six refer to the years from 1948 onwards, during which the Wood Sandpiper has occurred annually on autumn passage and occasionally during the spring. Of these records, fifty birds occurred during the month of August, twenty-seven in September, five each in May and July, and two or three each in April, June, and October.

It is a bird of lakesides and marshy pools rather than the tideline, and most of the Devon records have been from marshes bordering the estuaries, and from pools and reservoirs; twenty sightings come from the vicinity of the Exe estuary, including the Clyst and Exminster Marshes, ten from the Axe estuary, and the same number from the Taw estuary, including Braunton Marsh. About twelve have been observed on Lundy where it has occurred almost annually during recent years, and ten at Tamar Lake, where two remained for over a month during August and September 1949. Smaller numbers have been reported from the estuaries of the Otter, Teign, Erme, Yealm, Plym, Tavy, and Tamar; four each at Slapton Ley and the marshes at Bishops Tawton; and two occurred on the coastline at Wembury Point.

Most occurrences involve singles, but small parties of up to three birds have occurred on a number of occasions, and four, the largest number, were seen at Braunton Pill on the Taw estuary on 11 August 1967. This species, formerly confused with the Green Sandpiper, was first identified as a British bird by Colonel Montagu from one obtained on the south Devon coast.

COMMON SANDPIPER   *Tringa hypoleucos*

Summer visitor, passage migrant and winter visitor; breeds

The Common Sandpiper occurs plentifully on passage, frequenting the banks of reservoirs, streams, and lakesides in the spring, but keeping more to the tidal creeks and shore in the autumn. The spring birds arrive from mid-April onwards until well into May, and by early July the first autumn birds are back on the estuaries where they are plentiful, but well scattered, until the second half of September. On most of the Devon estuaries a few birds remain throughout the winter.

Formerly a regular breeding bird on Dartmoor, it has deserted many of its known breeding localities because of disturbance, but it seems almost certain that a few pairs still breed on one or two stretches of moorland rivers. Breeding was reported at Avon Dam in 1959 and 1962, and at a number of other localities during the 1950s. Cadover Bridge, formerly a regular site, was deserted in the early 1950s. It is doubtful whether any pairs breed on the Devon side of Exmoor, though it used formerly to nest on the Barle, just over the Somerset border.

Although normally occurring singly or in small parties, flocks of twenty or thirty and even fifty have occasionally been reported. It occurs regularly in small numbers on Lundy on both spring and autumn passage, but was more than usually numerous in 1959 when about thirty were seen together on 6 August.

REDSHANK   *Tringa totanus*

Resident and winter visitor, breeds

D'Urban stated that the Redshank, which was formerly abundant, still occurred in small flocks on migration and occasionally wintered. During the present century it has made a remarkable recovery and become an abundant passage migrant and common winter visitor on all the estuaries. It had also become established as a regular breeding species in several localities until the severe winter of 1962-3 not only exterminated the entire breeding stock but also about a quarter of the wintering population, and none has bred since.

The only breeding record known to D & M was at Slapton in 1894. It is recorded in *BB* 36 : 23 that a pair first nested at Braunton in 1908, following which it bred almost annually, with four pairs in 1960. Breeding on the Exe estuary was first proved in 1937 (*BB* 43 : 165), since when it became established and a few pairs nested regularly on Exminster, Powderham, and Clyst Marshes. Breeding on the Otter occurred during the 1940s and 1950s and on the Axe estuary from about 1957 to 1962.

It had probably attained its highest numbers by 1962, when totals of 1,000 were recorded on both the Tavy and Exe estuaries in August/September, with flocks of 100-200 occurring regularly on all the other estuaries during the previous decade. Although there has since been a continuing recovery, the numbers of passage migrants and winter visitors are still below the 1962 level.

The Redshank is unusual inland in Devon but has been observed at Tamar Lake and Burrator. Two examples considered to be of the Icelandic race (*T.t.robusta*) were ringed on the Exe estuary by David Cabot, one in January and the other in October 1962—*Devon Birds* 16:5 and 17:25.

Very small numbers occur fairly regularly on Lundy during spring and autumn migration.

SPOTTED REDSHANK   *Tringa erythropus*

Passage migrant and winter visitor

Regarded by D & M as a rare autumn straggler, of which they gave dated occurrences of only about sixteen individuals, the Spotted Redshank is nowadays a regular passage migrant and winter visitor which has been gradually increasing during the past thirty years, and has been recorded wintering annually since 1932. It now occurs regularly on the Exe, Axe, Tavy, and Taw estuaries and irregularly on all the other estuaries except the Dart. There are occasional inland records from Tamar Lake, and coastal occurrences at Prawle Point, Slapton, and elsewhere.

On the Exe it occurs during every month of the year, with autumn birds present from late July to the end of September, wintering birds from then until March, a few passage birds in April and May, and the occasional one in June. During the 1930s and 1940s the maximum did not exceed six, but since then the species has increased in numbers and become more widely dispersed, especially during the autumn. The peak numbers on the Exe in recent years have risen from nine in October 1952 to sixteen in September 1962 and nineteen in September 1966, while on the Tavy the species has become more frequent during the 1960s, with twelve in October 1961, twenty in September 1962, nineteen in September 1964, and up to about ten wintering.

Spring migrants, in their distinctive breeding plumage, are seen regularly in small numbers on the Exe during April or May, and irregularly elsewhere. Since 1954 the species has been recorded on Lundy on six occasions, involving ones and twos in the autumn and one individual in March.

The unusual influx in the autumn of 1962 involved more than

sixty birds in different localities, including nineteen on the Taw
estuary on 4 September.

GREATER YELLOW LEGS   *Tringa melanoleuca*
Vagrant

The only record of this rare vagrant from America is that reported
by Mrs F. E. Carter in *BB* 49:230 where a full account is given of
the occurrence. The bird in question was observed by Mrs Carter at
Tamar Lake on 11 October 1955. Feeding and preening at a small
pool beside the lake, its more robust appearance and stouter, up-
turned bill, compared with a Lesser Yellow Legs seen some time previously, were noted, as were its strident calls when it was later
flushed. The full and accurate description of this bird leaves no doubt
as to its correct identification.

LESSER YELLOW LEGS   *Tringa flavipes*
Vagrant

Until 1946, when one appeared on the Taw estuary, there was no
satisfactory Devon record of this trans-Atlantic drift migrant. The
bird mentioned by Ross in 1840 was considered by D & M to have
been incorrectly identified and was accordingly rejected by them.

On 8 January 1946, however, a bird of this species was observed
by Captain E. L. Shewell in a small creek which runs into the Taw
estuary near Fremington. It was later seen by a number of Devon
ornithologists and remained in the locality until 13 February, and
constitutes the only winter record; it is fully reported in *BB* 39:346.

In the autumn of 1954, following strong westerly winds in the
Atlantic, six Lesser Yellow Legs were recorded in various south coast
counties, including two birds in Devon. The first of these occurred
on the Clyst marshes at Topsham where I watched it almost daily
from 11 to 18 September, and it was also seen by F. R. Smith and
R. G. Adams. The other, which by its different breast markings was
judged to be a different bird, was watched at Ernesettle Marsh, Plymouth, from 24 to 30 September, being first seen by P. J. Dare and

later by others, of whom H. G. Hurrell obtained a colour film of it. Both of these occurrences are fully documented in the *Devon Report* for 1954 and are included in *BB* 48 : 363-6.

There have since been five further occurrences of this very graceful wader, all in the autumn, bringing the number to eight in all. Of these, two were observed in 1958, one being seen by R. G. Adams on the Exe estuary on 10 August and the other on the Kingsbridge estuary on 2 September by A. J. McVail. Two occurrences in 1959 relate to single birds on the Taw estuary on 30 August, and Lundy on the 26 to 28 October, the latter being the first record for the island. The last and most recent record is of a bird observed on the Braunton Pill on 24 and 25 August 1966 by A. J. Vickery and G. Jessup. These four mainland records were all accepted by the Rarities Committee and duly included in *BB*.

GREENSHANK *Tringa nebularia*

Passage migrant and winter visitor

The wild and noisy Greenshank nowadays occurs more regularly and in far greater numbers than were reported by D & M at the close of the last century. Indeed, they regarded it as being by no means common and referred to small flocks sometimes visiting the estuaries. Not only is it more numerous now, but small parties of up to about ten birds winter regularly on several of the estuaries, with up to twenty on the Exe. The returning birds begin to arrive in July, occasionally June, but it is September or October before the peak numbers are reached. Maximum counts for the Exe in recent years, where the numbers have markedly increased, were about 100 in September 1962, sixty in October 1963, ninety in September 1965, and seventy in the autumns of 1966 and 1967. Smaller numbers of up to twenty or twenty-five visit all the other estuaries, and odd birds are quite often recorded at Tamar Lake and Slapton, and on the Dartmoor reservoirs, during the autumn passage.

There is no marked spring passage but occasional birds are noted during April and May. Mostly singles but occasionally up to four birds are recorded annually on Lundy during the autumn and irregularly in the spring.

Page 137:
Woodlark at nest.
A local species which
has decreased during
the present century
but is still thinly
distributed throughout
Devon

Page 138:
Slapton Ley.
A Nature Reserve
owned by the
Herbert Whitley
Trust, this
freshwater lagoon
supports many
breeding
Reed, Sedge and
other Warblers;
it is visited by
many migrants
and is a winter
resort of a
variety of wildfowl

KNOT   *Calidris canutus*

Winter visitor and passage migrant

The Knot, nowadays a common winter visitor and passage migrant, has increased in numbers since the early 1950s. D & M regarded it as principally a passage migrant, and considered it was being badly reduced by shooting.

During the 1930s and 1940s the numbers on the Exe estuary, its main wintering ground, were well below fifty, and the Knot occurred mainly on autumn passage, with records of 150 in September 1934 and 118 in September 1949, but a flock of nearly 200 wintering birds occurred in February 1941. In 1950 the wintering flock amounted to 172 in February, since when the number gradually increased to 440 in February 1956, 621 in December 1960, and 900 in November 1962, but has since fluctuated between about 500 and 800. With passage migrants and small numbers of non-breeding birds, the species occurs on the Exe during every month.

From 1964 onwards 100-400 have wintered irregularly on the Plym estuary and the Hamoaze, and flocks of up to about thirty winter regularly on the Taw and Torridge, where there were 250 during the exceptionally cold February 1963.

Passage migrants in small parties or flocks of up to about twenty are widespread on estuaries and other coastal districts, including Slapton, Prawle Point, and Wembury, from about mid-August to mid-September and much less commonly during April and May. They have occasionally been recorded inland at Tamar Lake, where six occurred on 29 August 1962. On Lundy, ones and twos have been observed in April, August, or September on about nine occasions since 1947.

PURPLE SANDPIPER   *Calidris maritima*

Winter visitor

A bird of the reefs and rocky shores, the Purple Sandpiper rarely occurs elsewhere and never inland; it spends the winter on its chosen stretch of coast from its arrival in late October or early November until its departure in April. In some years a small passage involving only a few birds is detected in May.

J

On the south coast it occurs annually at Budleigh Salterton, Orcombe Point, Hope's Nose, Paignton, Prawle Point, and Wembury, while on the north coast it is regular at Ilfracombe, Woolacombe, and probably Croyde; but there must be many other, often inaccessible, localities where it occurs.

At most points the flocks average about ten in number, but more are recorded at Orcombe Point, Hope's Nose, and Wembury, where they vary from ten to twenty. Counts of more than usual include twenty at Orcombe Point in December 1949 and December 1953, and twenty-one in March 1954; twenty-five at Hope's Nose on 25 April 1960; and twenty-five at Wembury in February 1961, twenty-six in March 1963, and thirty-four in November 1966.

It is infrequent on Lundy where it has been recorded in only 9 of the 20 years from 1947, usually singly, but with a maximum of four passage birds in May 1949. There is nothing to indicate any change of status since 1900.

LITTLE STINT   *Calidris minuta*

Passage migrant

Although regular on autumn passage from August to October, mainly September, the Little Stint is much less regular in spring and in far smaller numbers, and it occasionally winters. Small parties of up to four or five visit the estuaries and occur annually on the Axe, Exe, and Taw, and at Wembury Point. The numbers vary from year to year, and while in some years the species may be reported from many localities, in others they occur only at one or two.

This minute wader occurred in unusual numbers in the autumn of 1960 when over ninety were reported in different parts of the county, with a maximum of forty on the Exe on 2 October and twenty-five at Westward Ho! on 4 October. The numbers had been building up from the end of August when the first birds arrived, and during this period seven were seen on the Axe, four at Plymouth, five on the Braunton Pill, and smaller numbers at several other localities.

Unusual records for other years include : twelve at Wembury on 19 September 1936 and ten to fifteen at Axmouth in October of that year (*BB* 30:230); fifteen on the Clyst on 29 September and twenty on the Taw on 28 September 1957; and six on the Exe on 17 May 1965 and eight at Wembury on 22 September 1967. The very few inland occurrences include two records at Tamar Lake and one at Wistlandpound, all of ones and twos. Single wintering birds have been recorded in about five winters, mainly on the Exe.

The Little Stint is rarely observed on Lundy, where the first fully authenticated record was of singles in August and September 1956. Eight were reported on 26 September 1957 and up to three in four subsequent years.

LEAST SANDPIPER   *Calidris minutilla*

Vagrant

Of the approximately twelve British records of this minute American wader, five have occurred in Devon. D & M knew of two occurrences for the county, both of single birds shot on Northam Burrows on the Taw estuary—the first on 22 September 1869 and the second on 22 August 1892. Although the *Ilfracombe Fauna and Flora* states that a bird was shot on the Taw by R. Riccard in 1890, this record is not mentioned by D & M and cannot now be substantiated.

Whilst there is no record of this straggler on the mainland during the present century, there are three recent occurrences for Lundy. A single bird, which arrived there soon after a hurricane in the Atlantic, stayed on the island from 24 to 26 September 1957. It was watched by Barbara Whitaker and A. J. Vickery at ranges down to 5 yd, and on one of the days was accompanied by eight Little Stints, with which a careful comparison was made. The occurrence is narrated in detail in the *Lundy Report* for 1957. The second occurrence for Lundy was on 8 September 1966, when one was mist-netted, together with a Semipalmated Sandpiper, by C. S. Waller and J. A. Ginnever, whose photographs and detailed descriptions of both species are contained in the *Lundy Report* for 1965-6. The same *Report* also lists the third example, which was mist-netted by C. S. Waller on 14 September 1966. These two records of Least Sandpipers were accepted by the Rarities Committee and included in their report in *BB* 60:318.

TEMMINCK'S STINT   *Calidris temminckii*

Vagrant

It is rather surprising that the records of Temminck's Stint in Devon are fewer than those of the Least Sandpiper. Of the three occurrences listed by D & M the first relates to a bird obtained on a salt marsh on the south coast in one November around 1800 and brought to Colonel Montagu, who described it in his *Ornithological Dictionary* under the name of the Little Sandpiper. The other occurrence is of two birds which were shot near Stonehouse Bridge, Plymouth, on 28 June 1837.

The only record of Temminck's Stint in the present century is of one observed at a freshwater pool on Braunton Burrows on 27 May 1961 by G. A. Sutton, whose convincing description of the bird and its calls is given in the *Devon Report* for 1961.

WHITE-RUMPED SANDPIPER   *Calidris fuscicollis*

Vagrant

Another vagrant from North America is the White-rumped Sandpiper, formerly known as Bonaparte's Sandpiper after the Italian ornithologist, Prince Charles Bonaparte. D & M admit four examples of this small Dunlin-like wader, all of which were shot on the Taw estuary near Instow in November 1870, at which time others were obtained elsewhere in Britain.

There have been three further occurrences, during the present century, all of single birds; namely, one carefully observed by O. D. Hunt and others at Wembury on 26 October 1952 and fully reported *BB* 46:261; another at the identical spot at Wembury from 12 to 22 October 1958, which was first seen by R. M. Curber and later by a number of observers, and described in detail in the *Devon Report* for 1958; and, lastly, one seen on Northam Burrows by D. F. Musson on 2 October 1961 and recorded in the *Devon Report* for that year.

With regard to the Wembury occurrence in 1952, it is of interest to note that the date coincided with the appearance of an American Robin on Lundy, and the disastrous 'wreck' of Leach's Petrels. As will be seen, all the Devon occurrences of the White-rumped Sandpiper, amounting to seven birds, were reported in the months of October and November.

PECTORAL SANDPIPER *Calidris melanotos*

Vagrant

Until 1950 the only authentic record of this trans-Atlantic drift migrant in Devon was that quoted by D & M of two birds, a male and female, which were both shot at a freshwater pool on Braunton Marsh on 12 September 1871. These authors also report a rather unsatisfactory record of 'some mentioned by Bellamy as shot on the Tamar' probably about 1839. In the years from 1950 onwards there have been fourteen occurrences, all but two of single birds, and all in late August, September, and October, except for one in February. Of these records, four relate to Lundy and the remainder to the estuaries of the south coast and Tamar Lake in north-west Devon. The occurrences, all of which are specified in the *Devon* and *Lundy Reports*, are as follows:

1950   Wembury, 7-15 October (P. J. Dare and H. G. Hurrell)
1950   Lundy, 12 October (D. Lee)
1951   River Clyst, Topsham, 11 February (R. F. Moore and S. C. A. Hunt)
1951   Exe estuary, 15-19 September (R. G. Adams and R. F. Moore)
1955   River Clyst, Topsham, 30 August to 5 September (R. F. Moore and F. R. Smith)
1955   Tamar Lake, two from 15-24 September (F. E. Carter)
1956   River Clyst, Topsham, 6 September (P. W. Ellicott)
1956   Ernesettle Marsh, Plymouth, 21-23 October (A. M. Common)
1960   Lundy, 10 October (LBO)
1961   Lundy, 31 August to 2 September (D. B. Iles)
1962   Tamar Lake, 29 August (SVB)
1962   Lundy, 2-8 September (LBO)
1963   Axe estuary, two on 12 October (P. A. Hill)
1967   Lundy, 24 September (LBO)

The Pectoral Sandpiper, designated in *The Handbook* as a vagrant, is the most frequent of the American waders in Britain.

DUNLIN   *Calidris alpina*

Resident, passage migrant and winter visitor; breeds

Principally a winter visitor and passage migrant, the Dunlin is also a regular breeding species, a few pairs nesting annually at about 1,900 ft in the Cranmere area of Dartmoor. Although D & M knew of no definite breeding records, they considered it probably bred on Dartmoor and quoted Dr E. Moore who stated in 1837 that a nest had been found there.

Despite repeated attempts during the 1930s and 1940s to obtain proof of nesting, and although breeding behaviour was observed in a number of years, it was not until 24 June 1956 that a nest with eggs was actually found, by A. V. Smith at Taw Head. A further nest with young and an egg was found in the same area by R. M. Curber and P. J. Dare on 20 June 1960, since when up to four pairs have been observed most years. It is probable that breeding pairs are fairly well dispersed in the centre of the moor, which is deserted after the breeding season.

On the estuaries the Dunlin is by far the most abundant wader, both as a winter visitor and passage migrant, and occurs in varying numbers throughout the year. The number wintering on the Exe has increased from about 3,000 in the 1940s to 8,000 during the 1960s, with a maximum of about 10,000 in November 1960, the peak occurring between November and February.

The numbers on spring passage in late April or early May show great fluctuations, from 100 or so in some years to a record of 4,000 in April/May 1967, and 2,000 in several recent years. Autumn passage occurs from July to September, with the main flocks in late August and early September. The same movements, but involving far smaller numbers, are recorded on all other estuaries, and small flocks occur regularly on the coast at Slapton, Wembury, and elsewhere. Except for those breeding on Dartmoor, only occasional Dunlin are recorded inland.

A pair was reported by Loyd to have nested on Lundy in 1904, apart from which it occurs there as a regular passage migrant. The most recorded since 1947 was twenty-four on 11 May 1950. As on the Exe, both the Northern *(C.a.alpina)* and Southern forms *(C.a.schinzii)* occur.

CURLEW SANDPIPER *Calidris ferruginea*
Passage migrant

On the Exe estuary the Curlew Sandpiper occurs annually on autumn passage, less frequently on the Taw and Torridge, and irregularly on the other estuaries. It is rare in spring, when just the occasional bird is reported during May in some years. The autumn passage occurs during late August to early October, with the main movement in mid-September. The numbers vary from year to year and are generally under about a dozen, but in some years many more occur. Peak numbers of from twelve to sixteen were recorded on the Exe estuary in the Septembers of 1936, 1945, 1949, 1952, 1956, 1957, and 1963. The largest flock recorded during the present century numbered seventy-three on 6 September 1959, but thirty-six were counted on 5 September 1954, and twenty-nine in the Septembers of both 1946 and 1953, all on the Exe.

Twenty were reported on the Taw estuary in September 1946 and small parties of about ten have been seen on the Avon, Tavy, Erme, and Otter during this century. Other than on estuaries, occasional birds have been seen at Wembury Point and the species has once been encountered at Tamar Lake. Davis mentions three records for Lundy up to 1939, since when ones or twos have been reported on three occasions.

D & M mention flocks of over a hundred occurring on the Taw estuary and they regarded the species as being second only to the Dunlin in numbers during autumn; they also refer to an instance when about 1,000 were said to have been present.

SEMIPALMATED SANDPIPER *Calidris pusilla*
Vagrant

Another trans-Atlantic drift migrant, this American wader has been recorded only once in Devon. A single bird was mist-netted on Lundy on 8 September 1966 by C. S. Waller and J. A. Ginnever. Their full description and photographs of this bird, showing its diagnostic, partially webbed toes, are given in the *Lundy Report* for 1965-6, and the record was accepted by the Rarities Committee, *BB* 60:319. It is particularly interesting that this bird was accompanied by a Least Sandpiper, and both were netted at the same time.

SANDERLING   *Calidris alba*

Passage migrant, winter visitor and non-breeding summer visitor

The vivacious Sanderling, although principally a spring and autumn passage migrant, occurs also as a winter visitor and non-breeding summer visitor, with varying numbers nowadays present throughout the year in the most favoured localities. D & M who regarded it as an irregular winter visitor, noted that it sometimes occurred in vast flocks.

A bird of sandy shores, it occurs in the greatest numbers on the Exe estuary in the south, and the Taw and Torridge, including Saunton Sands, in the north, and in small numbers at many other points along the south coast. It is rare inland, but very occasional birds have been recorded at Tamar Lake in the autumn. The largest flocks are recorded as a rule in May, when spring passage birds, many in breeding plumage, and often accompanied by Dunlins, occur on the Exe, some years in hundreds, but other years only in tens. An unusually heavy passage of 600 was recorded on the Exe on 19 May 1956, over 300 occurred on 22 May 1955, 200-300 on 23 May 1954 and 21 May 1961, and 100-200 in a number of recent years.

About 30-50 non-breeding birds are present on the Exe during June (200 on 14 June 1953), and up to 150 occur on autumn passage during July and August. Wintering birds on this estuary number 50-100, and there are about the same on the Taw, where over 120 were recorded in February 1945, and up to 100 on autumn passage.

Small parties are recorded annually at Wembury and not infrequently on the Avon estuary, at Slapton and Thurlestone, and occasionally at Woolacombe and on the Axe estuary. Ones and twos are reported irregularly on Lundy during August and September, and several occurred in May 1934.

BUFF-BREASTED SANDPIPER   *Tryngites subruficollis*

Vagrant

Another drift migrant from arctic America, the Buff-breasted Sandpiper has been reliably recorded on three occasions, while two other old records, not mentioned in *The Handbook*, must now be regarded as unproven. The three substantiated occurrences all refer to Lundy, where the first bird was shot by S. de B. Heaven in the autumn of 1858 and went to the Taunton Museum. The second bird was mist-netted on 24 September 1959, being only the second individual of this species to be ringed in Britain at that date. Two of these sand-pipers, which were identified by J. Martin and John Ogilvie on 27 and 28 September 1965, were reported in *BB* 59 : 289.

D & M mention a specimen said by F. W. L. Ross to have been obtained on the Exe in August 1851, but which was missing when his collection was handed over to the Exeter Museum in 1865, so could not be substantiated. The *Ilfracombe Fauna & Flora* contains the record of a bird, identified as this species, which was shot at Braunton by Major H. C. Chichester in 1908, but this record, like the preceding, cannot now be regarded as authentic.

RUFF *Philomachus pugnax*
Winter visitor and passage migrant

Until the 1950s the Ruff was regarded primarily as an autumn pas-sage migrant, but since then it has become established as a regular winter visitor, mainly on the marshes bordering the Axe, Exe, and Taw estuaries. Whereas D & M listed it as occurring only casually, it has been recorded annually since the 1930s and in gradually in-creasing numbers.

Up to 1948 the records related to small parties of usually under ten occurring on the south coast estuaries during August and Septem-ber and very occasionally in May. The first wintering birds on the Exe were a party of four in January 1949; in February 1954 twelve were seen on Exminster Marshes, and fifteen in December 1955; and twenty appeared at Ashcombe on 6 March 1955. The number on the Exe gradually increased to a maximum of forty-two in January/February 1962, then 20-30 until 1967, when fifty-one were present in January/February.

Up to about sixteen now winter on the Axe estuary and around five on the Taw.

Increasing numbers are also being recorded on spring and autumn passage, with a total of fifty-six in various localities, including Lundy, during September 1958. In 1965 twenty-two were seen on the Axe marshes in April and in the same month thirteen on the Plym; and in

1966 Ruffs were recorded in every month. The only inland locality where they occur fairly regularly in spring and autumn is Tamar Lake.

The first record for Lundy was in September 1937, since when five other singles have been reported, and a party of nine which occurred in September 1958.

AVOCET *Recurvirostra avosetta*

Winter visitor

The return of the Avocet as a regular winter visitor to Devon coincides almost exactly with its return to Suffolk as a breeding bird, while the gradual increase in the number wintering is related to the expansion of the Suffolk colony. From 1888 to 1924 not a single example was recorded in Devon, and it was not until the late 1940s that the Avocet was established as a regular winter visitor.

Four were recorded on the Exe in September 1924, followed by four records of singles, also on the Exe, during the years up to 1944. From 1947 onwards it has been recorded annually, the first two on the Tamar being observed in February 1948, since when it has wintered annually on a tidal stretch of this river between Warren Point and Halton Quay, and gradually increased. The maximum there in January 1952 was fourteen; twenty-five in January 1955; forty-six in December 1957; fifty-three in January 1962; sixty in December 1963; and since then it has fluctuated around the fifty mark.

Meanwhile, varying numbers of up to about ten have occurred annually on the Exe, some wintering but others on passage; six wintered on the Teign in 1960 and passage birds have been recorded on the Taw, Dart, Avon, and Kingsbridge estuaries, Slapton Ley, and one inland at Burrator in November 1965. Those wintering on the Tamar begin to arrive in late October and reach their maximum in late December or early January. They start to leave in late February and all have usually gone by mid-March. There is no record for Lundy.

A bird observed at Dawlish Warren in January 1961 was thought by the position of a ring above the tarsal joint of the left leg to have been ringed at Texel, Holland.

BLACK-WINGED STILT *Himantopus himantopus*
Vagrant

In 1830 Dr Edward Moore listed two undated occurrences in Devon of this rare vagrant from southern Europe. One was said to have been shot at Slapton Ley and the other obtained elsewhere in the county. D'Urban in the *VCH* said that no further example had occurred in Devon within the previous 60-70 years, and it was not until 1915 that the species was again reported. In that year, as related in *BB* 9:215, a single bird was clearly seen in an unspecified locality in north Devon on 6 November by F. B. Hinchcliff, who flushed the bird four times and had good views of it.

The next instance occurred on 17 September 1935 when, following a severe westerly gale, two immature Black-winged Stilts were observed by O. D. Hunt on the Yealm estuary at Newton Ferrers. The third occurrence for the present century concerns a flock of ten stilts seen by J. S. Jones in a cove near Bovisand on 10 May 1945 and reported in *BB* 38:337. The next occurrence of this most elegant wading bird was on 6 May 1949, when F. R. Smith observed four on a shingle bed in the River Exe, just below Countess Wear. They were seen later in the same day by R. G. Adams on the River Clyst at Topsham. The most recent record relates to a party of five seen by A. Gilpin on the Exe estuary on 12 September 1967.

GREY PHALAROPE *Phalaropus fulicarius*
Passage migrant

The Grey Phalarope, normally an uncommon autumn passage migrant, has been recorded in all but 5 of the 35 years 1933-67. Excluding the autumn of 1960, when most exceptional numbers occurred, the records involve approximately 145 individuals, of which about forty were observed in September, sixty in October, forty in November, and a few each in August, December, January, and March. Except for the occasional gale-driven bird found inland, all occurred on the coasts and estuaries, the majority in the south.

Small numbers of up to ten were reported in each of twenty-six of these years, but about twenty-seven occurred in 1952, fifteen in 1961, and sixteen in 1963, the records being spread fairly evenly along the south coast, and including about a dozen for Lundy.

All other records, however, are dwarfed by the great numbers that occurred on the south coast during the first week of October 1960, after cyclonic weather. On 5 October F. R. Smith observed at least 700 in Torbay off Hope's Nose, the sea being covered with them in flocks of up to 100. Of the great number of observations at many points on the entire south coast, the main numbers were twenty-five at Budleigh Salterton on 5 October, 108 at Dawlish Warren on 9 October, several dozen at Berry Head on 15 October, thirty-seven in Start Bay on 2 October, and seventy-one on 15 October. A few remained for several weeks and the last bird was seen on Dawlish Warren on 4 December.

The total number recorded from all Devon localities was about 1,200, though there may have been duplications in the moving flocks, which also included a small number of Red-necked Phalaropes. In *BB* 53 : 529-31 it is stated of this remarkable influx that at least 1,000 occurred off the Isles of Scilly on 15 October.

D & M mention seven particular years during the course of the nineteenth century in which more than usual occurred.

RED-NECKED PHALAROPE *Phalaropus lobatus*
Vagrant

Although the Red-necked Phalarope breeds abundantly in Iceland, its migration routes to and from there are evidently very wide of our shores, for the species occurs only occasionally on the Devon coast. D & M knew of only two examples obtained in the county, both from Plymouth—an immature bird in 1831 and an adult in summer plumage which was killed on the Hamoaze on the unusual date of 7 June 1869.

The dozen records for the present century mostly refer to single birds, and all occurred in the autumn months of September to November, being fairly evenly divided between the north and south coasts. The first bird was seen on the sea at Ilfracombe by A. S. Cutcliffe on 14 November 1946. The second was a dead bird, picked up at Dartmouth on 8 October 1955 and identified by M. R. Edmonds. In 1957 single birds were observed at the mouth of the Tamar and off Plymouth Hoe on 15 September and 3 November respectively, and two together were watched at a distance of only a few yards, on the Taw estuary, by Mrs D. Wilson and Dr Rogerson on 14 September, followed by two more, thought to be different individuals, by R. C. Stone at Fremington on 16 September. Detailed accounts of all these occurrences are included in the *Devon Reports* for the years stated. A single bird which was seen by B. G. Lampard-Vachell at Westward Ho! on 5 November 1958 is briefly mentioned in the *Report* for that year.

During the autumn of 1960, which was remarkable for its unprecedented numbers of Grey Phalaropes, two occurrences of Red-necked were reported. They comprise a party of three birds, amongst Grey Phalaropes, seen in Start Bay on 2 October by M. J. McVail, who recorded the facts in the *Devon Report* for 1960, and a single bird at Lundy on 18 October. The latter is the second record for Lundy, where the first, also a single bird, was seen on 11 November 1955. The most recent record for the county refers to a bird of this species which R. F. Coomber located at Northam Burrows on 6 September 1963, and adequately described in the *Devon Report*.

STONE CURLEW   *Burhinus oedicnemus*

Vagrant

This species according to D & M was a casual visitor, usually in the autumn and winter. The occurrences which they cited show that the now rare Stone Curlew formerly wintered in the county in very small numbers. That it has long since ceased to do so is evident from the complete lack of winter records during the present century.

Of the twelve records since 1900, involving thirteen individuals, all relate to spring and autumn passage, and are as follows: 26 April 1929 Dawlish Warren, 31 October 1931 Otter estuary, 27 September 1937 Exe estuary two, 15 March 1938 Lundy, 8 July 1938 Lympstone, 20 May and 19 October 1939 Lundy, 19 April 1942 Sidbury, 26 August 1944 Great Hangman, Combe Martin, 4 April 1947 Wembury, 5 September 1958 Cut Hill, Dartmoor, and 10 April 1963 Colyford Marshes. Most of these birds occurred in the south, the exceptions being one on Dartmoor, one in north Devon, and three on Lundy.

The Stone Curlew just survives as a breeding species in the neighbouring county of Dorset but, as in Devon, it rarely occurs in Somerset and Cornwall.

PRATINCOLE   *Glareola pratincola*

Vagrant

In the absence of any actual specimens obtained in the county, the two or three sight-records of the Pratincole listed by D & M were square bracketed. At least one of these records was subsequently accepted, as the species was accorded a full place in D'Urban's list contained in the *VCH*. The description of two birds seen on Dawlish Warren on 7 September 1851 sounds perfectly convincing.

For the present century there are two or three certain records and about the same number of probable occurrences of this Mediterranean species in Devon. The first entirely satisfactory record refers to an adult seen on Lundy on 21 February and again on 14 March 1945 by C. Robertson and reported in *BB* 39 : 93. In the same year one was seen by E. L. Shewell at Morte Point on 12 September and is described in the *Devon Report* for 1945. One occurred on Braunton Great Field from 15 to 18 May 1956, and was watched by T. G. Coward and I. W. Cameron, whose detailed notes are given in the *Devon Report* for that year. A record contained in the *Lundy Report* for 1962, relating to a bird seen on 26 April, although most probably correct, was not accepted by the Rarities Committee. Another record claimed as a probable Pratincole, but which appears to be absolutely correct, concerns a bird seen at Newton Ferrers on 19 May 1950, and is detailed in the *Devon Report* for 1950.

CREAM-COLOURED COURSER   *Cursorius cursor*

Vagrant

A vagrant from the deserts of Asia and North Africa, the Cream-coloured Courser has been recorded in Devon on three occasions, two of which are detailed by D & M. The first of their records involves a bird that was shot at Braunton at the end of October 1856, and another that was obtained a few days later; and their second record concerns two birds seen on Braunton Burrows in March 1860.

The third and most recent record is of a bird on Dawlish Warren from 11 to 14 October 1959 that was very closely watched by several different observers and described in detail in the *Devon Report* for 1959.

ARCTIC SKUA *Stercorarius parasiticus*

Passage migrant

A regular passage migrant, the Arctic Skua occurs annually along the south coast during the autumn, but is scarce and irregular in spring. Of over 400 recorded from 1935 to 1967, twenty-one occurred in May, thirty-three in July, eighty-four in August, 156 in September, ninety-five in October, thirteen in November, and up to four in all other months except March. (These are minimum figures as many singles are not detailed.)

Although this species is reported from many south coast localities, it is observed more frequently off the Exe than anywhere else, as skuas are attracted by the flocks of terns there in the summer and early autumn. Records for the north coast are few and irregular, and except for a bird picked up dead at Scorhill on the edge of Dartmoor in September 1959 there appear to be no inland occurrences.

Records of more than the usual singles or small parties include eight at Dawlish Warren on 11 September 1954; thirty-two seen between the Exe and Torbay on 6 September 1964; eight on the Exe on 3 August and eleven on 26 August 1965; eight there on 8 May, seventeen on 9 August, and fifteen on 4 September 1966; plus twenty-one seen off Slapton, fifteen off Start Point, and five off Stoke Point on 7 October 1967.

This species occurs irregularly at Lundy, where about a dozen, mostly singles, have been recorded since 1947, mainly in the autumn. Although the records for Devon suggest an increase since D & M's time, this may be due to the greater number of observers now.

GREAT SKUA *Stercorarius skua*

Irregular visitor

D & M regarded the Great Skua as a rare bird on the Devon coast and were able to list only about a dozen dated occurrences.

The *Devon Reports* contain about forty records involving some sixty birds, mostly singles but including two records of six and one of five; several of the birds were picked up dead or exhausted and many appeared during gales. Roughly one third were recorded on the north coast and two thirds on the south, with none inland. The two occurrences of six were off Combe Martin on 24 September 1964 and off Prawle Point on 8 September 1965, the latter during the course of a day of gales; and that of five was off Hope's Nose on 31 August 1960. Great Skuas were observed in 20 of the 40 years from 1928 to 1967.

The *Lundy Reports* contain one record, that of a single bird seen between the mainland and Lundy on 28 August 1948.

Eagle Clarke observed this species from the Eddystone Light in the autumn of 1901, and from observations made from research vessels, reported by G. M. Spooner (*Devon Report 1949*), it seems that this skua occurs regularly in the Channel during the autumn.

POMARINE SKUA   *Stercorarius pomarinus*

Irregular visitor

In view of the paucity of records of the Pomarine Skua during the present century, it is worthwhile quoting D'Urban's account of it in the *VCH*, in which he describes the species as : 'a passing visitor in spring and autumn, being most numerous at the latter season, when large flights appear off the south coast at intervals of about ten years. . . . This species is the least rare of all the skuas, and is generally to be met with in Torbay and Plymouth Sound in October, and sometimes in winter, but is not so numerous on the north coast of the county. It was very abundant in September and October 1901 off the Eddystone Lighthouse, according to Mr Eagle Clarke'. That there can have been no likelihood of wrong identification is apparent from the number of specimens shot at various points along both coasts, particularly during the autumn of 1879.

The position is very different at the present time, for the Pomarine Skua is now, except for the Long-tailed, the rarest of the four species that visit Devon, and is little more than a vagrant.

In addition to the twenty or more examples listed in the *Devon Reports* from 1937 onwards, two records for the present century are contained in D'Urban's notebook—one relating to an example shot outside Salcombe on 1 January 1903, and the other record being a note by E. A. S. Elliot saying that skuas, mostly Pomarine, were in great abundance off the Kingsbridge estuary on 16 May 1918. A further record, mentioned in the *Ilfracombe Fauna & Flora*, states

Page 156: Burrator reservoir Dartmoor. The reservoir and its surrounding coniferous woodlands have provided habitats and sanctuary for a variety of species which are absent from open moorland

that one obtained in north Devon in 1912 was preserved by J. Rowe, the taxidermist at Barnstaple.

The recent occurrences, taken from the *Devon Reports*, relate to an immature bird seen at close quarters on the Exe estuary by R. G. Adams on 3 January 1937, an adult observed at Dawlish Warren on 22 August 1944, and another sight-record of one seen at the same place on 26 November 1945. The record of one seen off Ilfracombe on the very unusual date of 8 July 1947 lacks any supporting evidence. On 22 November of that year, however, a female was picked up in an exhausted condition at Slapton Ley by J. Barlee. On 9 September 1962 one was seen at close range at Dawlish Warren, where single birds were again seen on 1 and 20 September 1964. Four were identified here by R. Khan on 2 August 1965 and one on the following day, while one was seen off Combe Martin by R. F. Coomber on 3 September of the same year. R. Burridge reported five flying west past Stoke Point on 7 October 1967.

There is no record of the Pomarine Skua having occurred at Lundy, but doubtless is did during October and November 1879 when numbers were driven by gales on to the south-west coast of England and were reported in all four south-western counties.

LONG-TAILED SKUA *Stercorarius longicaudus*
Vagrant

This species, normally a very scarce and irregular visitor to the south-west, occurred in some numbers during October 1891 when severe gales drove many on to the shores of the English and Bristol Channels, to be ruthlessly killed and hoarded by the private collectors of those days. Previous to this there were only about four known occurrences of the Long-tailed Skua in Devon, all from the south coast. D & M, who recorded the 1891 occurrences in detail, listed sixteen birds which were obtained on this occasion, one at Ilfracombe, another on the Taw estuary at Barnstaple, two at Exmouth, one at Milton Ley, and no less than eleven of a flock that came into Bigbury Bay. These birds were mostly adults but included others in various intermediate plumages. At the same time other examples were taken on the adjacent coasts.

Since 1891 there have been only three further reports of this very

K

elegant skua in Devon. The first of these, an adult, was identified by
W. B. Alexander and M. C. Radford on 11 September 1942 as it flew
past the boat on which they were travelling to Lundy a mile or two
east of the island, and was recorded in *BB* 36 : 140. The second report,
which bears no supporting evidence, refers to a single bird said to
have been observed off Ilfracombe on 22 October 1949. The third, an
almost incredible record, relates to a bird seen by P. J. Dare, a most
meticulous observer, near Fernworthy on Dartmoor on 11 June 1961.
This bird, which was flying on a steady north-west course, is carefully described in the *Devon Report* for 1961.

IVORY GULL   *Pagophila eburnea*
Vagrant

The Ivory Gull, described as a graceful tern-like bird, which breeds
in the high Arctic and normally winters in the Arctic Ocean, has
been recorded only once in Devon. An example in nearly adult
plumage was shot at Torquay on 18 January 1853 and is still preserved in the Torquay Museum. Details of the occurrence were duly
recorded by D & M, who remark that the bird when first seen was
apparently in an exhausted condition. There has been no further
Devon record of this rare vagrant during the past hundred years and
more.

GREAT BLACK-BACKED GULL   *Larus marinus*
Resident and winter visitor, breeds

According to D & M this species was no longer breeding on the mainland at the close of last century, and occurred only in ones and twos,
mainly in winter, but a pair or so still bred on Lundy. Loyd considered a pair bred at Beer Head in 1926 and 1927.

   In the early 1930s T. H. Harrison and H. G. Hurrell (*BB* 25 : 136-8)
reported a great increase and stated that about thirty-one pairs bred
along the south coast and some on the north. During recent years
breeding has been reported from a number of localities along both
coasts, mostly on offshore stacks, including up to nine pairs on the
Oarstone at Torbay and up to six on the Mewstone at Wembury,

and at High Peak, Brandy Head, Ladram Bay, Bolt Head, and on the north coast at Woody Bay, Ilfracombe, Lee, Baggy Point, and elsewhere.

In winter flocks of 100 and more occur regularly on most south coast estuaries, while occasionally great numbers are reported, particularly after storms. Such records include: 2,000 on the Erme estuary in January 1950, 2,000 at Slapton in February 1953 and 1,250 there in December 1960, 2,000 on the Exe in December 1964, and 1,000 at Slapton in February 1966.

It occurs inland in all months, usually at rubbish dumps, and small numbers have been reported from various localities on Dartmoor, particularly from the reservoirs.

On Lundy the breeding population had increased to sixteen pairs in 1923 and Perry gave the number as about fifty-seven pairs in 1939, while from 1947 to 1966 it has varied between thirty and forty-seven pairs and tends to increase despite some measure of control.

LESSER BLACK-BACKED GULL *Larus fuscus*
Summer visitor and passage migrant, breeds

This species occurs principally as a passage migrant; it breeds regularly on Lundy and in small numbers on the mainland coast, and a few winter. The spring passage occurs during February and March, the first birds usually being recorded on the coast and estuaries in about mid-February. The greatest numbers occur as a rule during March, when flocks of 100 are not infrequent, and one of 200 was reported at Slapton on 13 March 1960.

A small overland passage, chiefly in spring, has been detected from a number of coastal and inland localities, including central Dartmoor. Autumn passage is recorded, chiefly along the south coast, during September and October. Most years a few wintering birds are recorded.

It is scarce as a breeding bird on the mainland, but a few pairs nest on the north coast. It was regularly recorded at Baggy Point and between Ilfracombe and Lee during the 1930s to early 1950s, and may still occur. Nesting was reported near Hartland Point in 1943, and C. G. Manning informs me that two pairs now breed near Woody Bay. On the south coast three pairs nested at Stoke Beach in 1939, and about fourteen pairs were reported between Bolt Head and Bolt

Tail during the 1930s, where the number was stated as twelve pairs in 1946. In that year two pairs nested at East Prawle and a few may still nest in the Salcombe area.

The breeding population on Lundy was said by D & M to be small but was estimated by R. Perry as 350 pairs in 1939. It has since declined to less than 100 pairs in 1952 and around forty in 1957, but there has been some recovery during the 1960s, with seventy pairs in 1965 and a continued increase reported in 1966. The birds are present on the island from March until August, with a few stragglers in September and October, and occasional wintering birds.

The Scandinavian form (*L.f.fuscus*) occurs fairly regularly in Devon, on spring and autumn passage, but in very small numbers.

HERRING GULL   *Larus argentatus*
Resident, breeds

By far the commonest gull in Devon, the Herring Gull is abundant throughout the year along the entire coastline and is plentiful inland especially during the winter, when foraging flocks occur on arable and pasture land. A resident species, it is largely sedentary but there is some dispersal from the main nesting colonies after the breeding season. This is particularly so of Lundy, one of its principal breeding grounds, which is practically deserted during the winter, when the population drops to less than 100 birds. On Dartmoor also it is decidedly scarce during the winter. P. J. Dare's observations around Postbridge show that foraging parties frequent the cultivated parts of central Dartmoor principally during the months of April to August, the birds evidently roosting on the nearby reservoirs.

The many miles of precipitous cliffs along both the north and south coasts of Devon provide innumerable nesting sites for this gull, which breeds very plentifully along almost the entire coastline, with larger colonies at many of the headlands such as Baggy, Highveer, and Foreland Points in the north, and Beer, Berry Head, and Bolt Head in the south, to mention but a few. In one locality, at Ladram Bay, these gulls nest on the pebble beach, and in one or two places they have been recorded as nesting on chimney stacks, but as yet there are no reports of breeding inland.

No estimate has been made of the number of pairs breeding in

Devon, but it must run into many thousands. Attempts to count them on Lundy have been made by R. Perry, who gave the figure as 3,000 pairs in 1939, and H. J. Boyd who estimated it as 1,230 pairs in 1949. P. Davis mentions estimates of about 2,000 pairs in 1922 and 1923, made by Loyd. Boyd suggested that the large decrease since 1939, amounting to three fifths of the population, may have been due to the heavy toll of eggs taken for human consumption during these ten years, when communication with the mainland was very restricted. Sample counts taken in one corner of the island over a period of eight years from 1949 to 1956 fluctuated too widely to suggest any particular trends (*Lundy Report* for 1956 pp 28-30). The *Lundy Report* for 1965-6 gives the breeding population in 1966 as over 3,500 pairs, which appears to be at variance with the figure of 1,190 pairs counted in the previous year.

The BTO enquiry into the wintering of gulls in Britain showed that the only inland roost of Herring Gulls in Devon in January 1953 was that of some 3,000 birds at Tamar Lake. The number on the Exe estuary, a coastal roost, was estimated by F. R. Smith as 3,500 birds at that time. The morning flights inland from the coast, and back again in the evening, are a feature of some of the coastal districts during the winter, but they include Common and Black-headed as well as Herring Gulls. At Paignton on 3 February 1963 M. R. Edmonds estimated that 10,000 Herring Gulls, together with Common and Black-headed Gulls were feeding on vast quantities of dead shellfish cast up on the beach by persistent easterly winds.

H. J. Boyd, in the *Lundy Report* for 1956, states that the recovery of Herring Gulls ringed as nestlings on Lundy shows them to have been fairly evenly distributed on the north and south shores of the Bristol Channel, with the odd bird or two from further south.

Although D & M give no figures, but describe this species as abundant, it has evidently increased very considerably in Devon as a whole during the present century.

COMMON GULL *Larus canus*
Winter visitor

An abundant winter visitor, the Common Gull occurs on the estuaries, along the entire coast, and commonly inland. D'Urban in *VCH* erroneously suggested that a few pairs might breed on both the north and south coasts, but there is no evidence of it ever having bred in Devon.

The first returning birds arrive in July and August, and many remain until well into April, probably augmented by passage birds. Flocks of 100-500 are regularly reported roosting on the estuaries and moving inland by day to pasture fields and ploughland, but it is scarce on Dartmoor. Some sample figures from the *Devon Reports* include 800 on the Torridge above Bideford in November 1947, 1,200 on the Teign estuary on 28 March 1948, 200 at Tamar Lake in January 1952, 600 at Powderham in February 1957, 750 on the Yealm in January 1957, 1,000 on the Taw estuary in January 1958, 600 at Plymouth in March 1959, and 400 at Slapton in March 1960.

This gull occurs only occasionally on Lundy, where a few spring or autumn birds are recorded in some years but none in others, and a maximum number of about fifty were present on 28 November 1952.

GLAUCOUS GULL   *Larus hyperboreus*
Scarce and irregular winter visitor

Occurring slightly more often than the Iceland Gull, with which it is easily confused, the Glaucous Gull is also a scarce and irregular winter visitor. D & M stated that many occurred at Plymouth in 1872; other than this record, the occurrences now appear to be much the same as during the nineteenth century, with singles being reported from time to time on both coasts.

Excluding the individual that has wintered on the Exe estuary every year since 1961, about thirty examples have been recorded in Devon during the past 40 years, all in the months from December to May, and of that number eight occurred in January, six in February, and seven in April. Of these, three occurred on the north coast, one inland at Torrington during the extremely severe weather of February 1963, seven at Plymouth, a few each at Slapton and the Exe estuary, and the remainder elsewhere on the south coast.

The wintering bird on the Exe, which was first recorded in April 1961, was present continuously from September 1961 until March 1963, and thereafter during every winter until 1966-7, being last seen in March 1967. During this period it changed from immature to adult plumage by about 1963, and from 1963 regularly left the area towards the end of March and returned again in late July, August, or September.

There is no definite record of this species on Lundy, but two occurrences of white-winged gulls in 1949 and 1952 are indeterminate between this and the Iceland Gull.

ICELAND GULL   *Larus glaucoides*

Scarce and irregular winter visitor

A scarce and irregular winter visitor, the Iceland Gull occurs less frequently in Devon than the larger Glaucous Gull. D & M, who regarded it as a casual winter visitor, listed about thirteen occurrences of single birds, all but one of which were from the south coast. In addition to these, they remarked that it was numerous in the winter of 1874-5, but gave no actual numbers. The only record they quoted for the north coast was that of a bird shot on the Taw during January 1893.

Of the twenty or so records for the present century contained in the *Devon Reports*, all but one refer to single birds, the exception relating to an adult and an immature at Slapton on 25 March 1958. The majority of these birds occurred on the south coast, mainly at Plymouth, Slapton, and Dawlish Warren, but one was observed at Appledore on 24 January 1955, and another, an immature bird, was present on the Torridge estuary at Bideford from 16 February until 13 March 1962. One or two of the occurrences are of particular interest because they refer to occasions on which both the Iceland and Glaucous Gulls were seen at the same time. One such incident was on 7 May 1950 when O. D. and D. B. Hunt were able to observe and compare both species at close range on Plymouth Breakwater, with three other species of gull also present. Mr Hunt's detailed descriptions of both birds are given in his report in *BB* 43 : 409-10, in which it is remarked that the winter of 1949-50 was a particularly good one for both these species in Britain. Similarly, N. A. Wesley and A. C. Sawle watched both at Dawlish Warren on 9 March 1963.

The Iceland Gull has been recorded from Lundy on four definite occasions: 11 April 1939 (R. Perry), 5 April 1950, an adult on 28

March 1952, and an immature bird on 27 and 28 November 1954; while two occurrences of white-winged gulls were indeterminate between this and the Glaucous Gull. Whereas D & M's records for Devon were for the months from October to March and included only one for May, eight of those for the present century refer to April and May.

GREAT BLACK-HEADED GULL   *Larus ichthyaetus*
Vagrant

The first example of this Asiatic species to be recorded in the British Isles was shot at Exmouth at the end of May or beginning of June 1859 by William Pine, a boatman. It was presented to F. W. L. Ross of Topsham, whose collection was subsequently bequeathed to the Royal Albert Memorial Museum, Exeter, where the specimen may still be seen. Details of the occurrence, from Ross's original account, are given by D & M. It has been recorded in Britain on only a few subsequent occasions, but this is the one and only record for Devon.

MEDITERRANEAN GULL   *Larus melanocephalus*
Vagrant

This species is not mentioned by D & M, and the first record of its occurrence in Devon was not until 1964, since when it has been reported on twelve occasions during the four years to 1967.

The first record for Devon was that of an adult moulting from summer plumage, closely observed at Dawlish Warren on 28 and 29 July by G. Vaughan and P. W. Ellicott. In the same year a second bird in sub-adult plumage was seen on the Teign estuary near Shaldon Bridge by R. Khan on 23 September. Full accounts of both occurrences are contained in the *Devon Report* for 1964. In the following year, 1965, one in winter plumage was seen by many observers at Dawlish Warren from 7 to 31 January; another at Shaldon on 16 December (R. Khan); one at Chelson Meadow bordering the Plym estuary on 4 March (R. Burridge); another on the Plym estuary on 2 July (P. Harrison); and an immature bird at Wembury on 12 November (P. Harrison); these five records are briefly listed in the *Devon Report* for 1965. The species was again well recorded in 1966, with four occurrences relating to three separate sightings—at Dawlish Warren on 9 January, 9 August, and 5 November, and on the Plym estuary on 1 July, all of which are listed in the 1966 *Report*. In 1967 an adult frequented Dawlish Warren during July and August during which period it gradually assumed winter plumage. All these records appear to relate to birds wintering in the English Channel.

BONAPARTE'S GULL   *Larus philadelphia*
Vagrant

This rare straggler from North America has been reliably recorded once in the county, when an immature bird was observed feeding with other gulls at the sewage outfall at Sidmouth on 30 October 1961. It was seen and identified by I. C. T. Nisbet, who is thoroughly familiar with the species in the USA, and whose detailed account of the occurrence is included in the *Devon Report* for 1961. This record, which was accepted by the Rarities Committee, appeared in *BB* 55 : 574.

LITTLE GULL   *Larus minutus*
Passage migrant and winter visitor

A regular but uncommon passage migrant and winter visitor, the Little Gull has been recorded in 24 of the 30 years from 1938 to 1967 and annually since 1952. Of approximately 145 examples recorded during this period, fifteen occurred in March, twenty-two in September, fifty in October, and six to ten in all other months except May and June, which had three and one respectively.

About ninety Little Gulls were recorded on the Exe estuary, ten in Torbay, eight on the Taw and Torridge estuary, eight on the Otter, and around five each on the Axe estuary, Slapton Ley, Wembury Point, and Plymouth Sound, with two inland occurrences at Tamar Lake and one at Paignton reservoir. The great majority of these birds occurred singly, but some in small parties of up to four. An exceptional influx occurred in the autumn of 1961 when many were observed at Dawlish Warren between 7 and 9 October, with a peak of thirty-three on 8 October.

The three Lundy records of this very small gull comprise two singles encountered between there and the mainland in September 1948 and April 1966, and an old record of one in October 1891.

D & M, who regarded the Little Gull as a casual visitor to Devon, listed about thirty-two records for the nineteenth century.

BLACK-HEADED GULL   *Larus ridibundus*

Resident and winter visitor, breeds irregularly

Although mainly an abundant winter visitor, the Black-headed Gull is present throughout the year, as many non-breeding birds remain on the estuaries during April to June. Returning birds arrive from July to about November and the maximum numbers probably occur in mid-winter. It forages in great numbers inland, mainly on arable and grassland, returning in the evening to roost, often in vast numbers, on the estuaries. It also occurs commonly along both coasts, but is infrequent on Dartmoor.

The number roosting on the Exe estuary was estimated in the winter of 1952-3 to be around 28,000, with another 2,000 on the Teign, up to 5,000 on the Axe, and other roosts elsewhere, including an inland roost of over 700 on Tamar Lake.

This gull did not breed in Devon during the nineteenth century, but a small colony of up to twenty pairs existed at Braunton Marsh from about 1915 until 1961, when breeding was last recorded. The population in 1929 was about fifteen pairs; there were fifteen to twenty nests in 1946, and thirteen in 1948, after which the number declined until the last two nests were reported in 1961, following increased disturbance.

There are few numerical records of this extremely abundant species, but the *Devon Reports* mention counts of 5,000 on the Torridge in February 1946 and at Saunton in August 1959, and the non-breeding population on the Exe estuary was given as 400 in June 1947.

Small numbers occur annually on Lundy, where it has been observed in all months, but rarely more than twenty at once.

SABINE'S GULL   *Larus sabini*

Vagrant

In summarising the history of Sabine's Gull in Devon, D'Urban stated in *VCH* that less than a dozen examples, all in immature plumage, had occurred on the south coast, all in the month of October, and none had been met with on the north coast. In the 60 years since that was written, there have been six reliable records in the county of this most beautiful and graceful little gull, as *The Handbook* describes it. The first of these refers to an immature male which was shot on the Exe estuary by T. McClaughlin on 4 October 1920 and, as stated in *BB* 14:211, was presented to the Exeter Museum. The small 'invasion' of these gulls during September 1950, recorded in *BB* 44:254, included one Devonshire occurrence, namely an immature bird which was observed on the sands at Woolacombe by E. H. Lousley on 22 September. The next record is a particularly interesting one because it refers to an adult in summer plumage that occurred at Ernesettle, Plymouth on 22 July 1956, the bird being adequately described by M. A. Common, who mentions amongst other points the tern-like flight, striking wing pattern, forked tail, and dark grey head. In the autumn of 1958, when several examples of this Nearctic species occurred in the south west of England, an immature bird was reported from Lundy, the first and only record for the island, on 28 September. Two further records for the mainland concern single birds seen at Dawlish Warren on 9 October 1960 and 9 August 1966, the former coinciding with the occurrence of many hundreds of phalaropes on the Devon coast.

KITTIWAKE *Rissa tridactyla*

Resident, breeds

The Kittiwake, although principally a summer visitor and pelagic in winter, occurs in varying numbers on the coast in all months of the year, and especially during the autumn, when numbers are driven inshore by gales. It was described by D & M as the most numerous gull on the Devon coast, but at that time, except for an isolated record of a pair breeding at Berry Head in 1873, it nested regularly only on Lundy.

Breeding was not again recorded at Berry Head until 1930 when fifteen pairs nested. This colony increased to a maximum of 225 nests in 1941 (*BB* 35:86), and thereafter gradually decreased to forty pairs in 1952 and seven in 1962, but possibly increased in 1967. A fresh colony of sixty pairs discovered at Scabbacombe Head in 1948 decreased during the next twenty years to around twenty pairs in 1967. On the north side of Torbay a new colony meanwhile became established on the Lead Stone where thirty pairs bred in 1947, increasing and spreading to Hope's Nose where six pairs bred in 1951. In 1958 there were some 170 pairs at Hope's Nose and eighty on the adjacent Oar Stone, and this colony has continued to increase during the 1960s. In 1967 a small colony was found at Bolt Head.

On Lundy, which is occupied from March until mid-August, the breeding population was given by Perry as 3,000 pairs in 1939, since when it has decreased to 2,026 occupied nests in 1951, 1,308 in 1955, and only 760 breeding pairs in 1962; but it was stated that there were 1,225 pairs in 1965.

Coastal movements are recorded at Start and Prawle Points, mostly during September/October, and some 600 passed Berry Head in an hour on 15 September 1966. Occasional storm-driven birds have been reported inland on southern Dartmoor and elsewhere. More than usual were 'wrecked' by south-westerly gales in February 1957, with about 100 in Plymouth Sound.

BLACK TERN   *Chlidonias niger*

Passage migrant

D'Urban in the *VCH* summarised the status of the Black Tern as: 'A spring and autumn visitor, usually in small numbers, but at long intervals vast numbers have appeared at the latter season on our shores and estuaries', and he quoted 1849, 1859, 1866, and 1868 as years in which many were seen.

A regular passage migrant, the Black Tern nowadays occurs annually along the south coast in autumn, but is rather less frequent in the north where, in some years, none are recorded. It has been reported in the county in every year from 1934 to 1967, but is less regular on spring migration than in the autumn. The spring passage of this species normally passes well to the east of Devon, which receives only the odd individuals, or more rarely the fringe of an exceptional movement such as occurred during May 1954 when a flock of at least seventeen was observed on the Exe estuary on 9 May by F. R. and A. V. Smith and myself. This movement, which was concentrated in the Severn estuary, the Midlands, and East Anglia, and involved some 2,000 birds, is fully documented by R. F. Dickens in *BB* 48 : 148-69. At such times far more are recorded on the Somerset reservoirs than ever occur in Devon.

The spring migration, consisting of adults in breeding plumage, is confined almost entirely to the month of May, with very occasional birds in late April and early June, but the autumn passage is much more leisurely and extends from about mid-August, occasionally late July, until late October, with individual birds or small parties remaining in a favoured locality for days or even weeks. Some of the August birds are still in summer plumage but the majority of autumn migrants are in winter or immature plumage; and sometimes an adult is still accompanied by a bird of that year.

Although this most graceful bird occurs chiefly on the coast, particularly on the Exe estuary, Slapton Ley, and on the Taw estuary in the north, it has been recorded at inland waters on a number of occasions, odd birds having been seen in several years at Tamar Lake both in spring and autumn. There is also an interesting record, indicating overland passage, of a party of nine which B. Gooch witnessed flying upstream, the full length of Fernworthy reservoir, and continuing westwards out of sight on 22 September 1957. This water is just over 1,100 ft above sea level and a flight westwards would have taken the terns over the highest part of Dartmoor. There is also the record of a single bird, probably storm-driven, which was picked up in a field at Aish near South Brent on 27 October 1952.

The flocks or small parties which enter the estuaries often travel some miles inland, up the rivers, and may be seen feeding over flooded water meadows and small marshes, as, for instance, a flock of twenty-five observed by A. J. Vickery at Bishops Tawton on 25 September 1957, and a single bird seen by G. H. Gush on the Torridge at Weare Giffard on 26 August 1947. Whilst Black Terns may be encountered at almost any point along the south coast during the autumn, the most favoured localities are the Exe estuary and Slapton Ley, where singles and small parties of up to four or five are observed every autumn.

Some of the more unusual records for recent years include a flock of fifteen on the River Caen at Braunton on 9 September 1956; eight on the Exe estuary on 10 August 1958; fourteen at Slapton Ley on 1 May 1965; and up to fifty on the Taw estuary between 12 and 21 August of the same year, observed by J. Coleman Cooke and A. J. Vickery.

Although there is no record of this species for Lundy itself, the *Lundy Report* for 1956 mentions that three birds were encountered about 9 miles east of the island during a crossing from the mainland on 2 September of that year.

WHITE-WINGED BLACK TERN   *Chlidonias leucopterus*

Vagrant

D & M list four occurrences of this rare vagrant from south-eastern and eastern Europe. The only satisfactory record among these is of one shot in the harbour at Ilfracombe on 2 or 3 November 1870. The other three records are undated and refer to single birds, of which one is said to have been obtained at Kingsbridge, one from Plymouth Sound, and the last, a bird in the Exeter Museum, is presumed to have been obtained on the Exe estuary.

On 9 May 1954 I observed an adult in full summer plumage in company with a flock of seventeen Black Terns on the Exe estuary. Being in my boat at the time, I was able to watch it at very close range as it circled around me, feeding and settling on an exposed bank to which it kept returning to rest. It was also seen by F. R. Smith and on the following day by P. J. Dare, and is fully reported in *BB* 48:178. This occurrence coincided with a passage of Black Terns in various parts of England, but appears to have been the only example of the present species.

WHISKERED TERN   *Chlidonias hybrida*

Vagrant

The only authentic record of this rare vagrant to Devon is an adult in summer plumage which was picked up alive but exhausted by some fishermen off Plymouth on 10 May 1865. Details of the occurrence are given by D & M, and the record is quoted in *The Handbook*. The first example for the British Isles of this marsh tern was obtained just outside the Devon border at Lyme Regis in August 1836.

GULL-BILLED TERN   *Gelochelidon nilotica*
Vagrant

The Gull-billed Tern, a cosmopolitan species which in Europe breeds as close to Britain as Denmark and the Netherlands, is an extremely rare vagrant to Devon where it has been recorded only twice. D & M record that an immature bird was obtained near Plymouth in October 1866. After an interval of 101 years, a tern of this species was observed at Wembury Point on 5 September 1967 by R. Burridge, who noted the stout black bill which he described as looking heavy for such a graceful bird. This record was accepted by the Rarities Committee, *BB* 61 : 345. The Gull-billed Tern was first made known as a British bird by Colonel Montagu in 1813, from an example obtained in Sussex at an earlier date.

CASPIAN TERN   *Hydroprogne tschegrava*
Vagrant

The old reports of this rare vagrant were evidently not considered sufficiently reliable for inclusion in *The Handbook*, which contains no record for Devon. D & M list three occurrences : one said to have been shot on the Exe estuary near Topsham, an immature bird reported from Teignmouth in October 1861, and one said to have been shot in Torbay on 28 September 1873.

The inclusion of this very large tern in the county list was settled beyond doubt by the recent occurrence of one on the River Axe on 6 July 1966. The bird was closely watched by R. Cottrill, who noted its heavy build and huge bill, and whose accurate description of it is given in the *Devon Report* for 1966. This was one of ten or eleven occurrences in England during June, July, and August 1966, reported in *BB* 60 : 321.

COMMON TERN   *Sterna hirundo*
Passage migrant, has bred

The Common Tern, a regular passage migrant, occurs in varying numbers on the south coast from its arrival in mid-April until the last stragglers depart in mid-October. On the Exe estuary, where the greatest numbers gather, some are present in all months from April to October, but the main flocks congregate during July and August, though not nowadays in such numbers as were recorded during the 1930s and 1940s. In August 1936 the number was estimated as 1,700, with 2,000 in 1937; in August 1938 there were 1,000 reported; 1,100 in August 1942; and 2,000 in August 1945 and again in August 1950. During the following decade the peak numbers ranged from 200 to 700 : in September 1961 it was 1,500, but 800 in the autumn of 1964, with 100-300 in other recent years, apparently depending on the abundance of sand eels.

But for the increasing human disturbance there would almost certainly be a breeding colony on Dawlish Warren where four pairs attempted to nest in 1950, and one pair in 1954, 1957, and 1963, these being the only known nesting records for Devon.

Passage is recorded annually at other points along the south coast, notably at Hope's Nose and Slapton, where the numbers occasionally reach 100, and smaller parties occur on all the estuaries and sometimes several miles up the rivers. On the north coast the Common Tern visits the Taw and Torridge fairly regularly during the autumn, when 30-60 and occasionally more are observed, with a maximum of 200 on 13 August 1965.

This tern is infrequent inland, but parties of up to ten have been seen at Tamar Lake and singles at most of the inland waters.

Parties of up to about six are recorded at Lundy most years, usually in August and September. An unusually large flock of about 140 occurred there in bad weather on 5 October 1958.

ARCTIC TERN   *Sterna paradisea*

Passage migrant

Amongst the many Common Terns that occur on passage migration, a few definite Arctic Terns are recorded annually by experienced observers, often from amongst the mixed flocks that rest on the shore at Dawlish Warren, where they are easily approached. The remains of a number of terns found there in the autumn of 1961 were critically examined and four proved to be this species.

More records relate to the months from July to October than to

April and May, but most years ones and twos are recorded on spring passage and small parties of up to four or five in autumn, though these probably represent only a fraction of the actual number involved. More are recorded on the Exe than elsewhere, but Arctic Terns occur regularly along both the north and south coasts, where individuals are reported from many localities, including fairly regular records from the Taw and Torridge estuary, but there appear to be no definite inland occurrences.

D & M relate that great numbers were driven on to the south coast by adverse weather in May 1842, when hundreds were killed on the Exe.

There are no recent specific records of this species at Lundy, where the usual small flocks occurring in autumn are reported as Common or Arctic Terns, but Loyd states that an Arctic was identified by H. N. Joy between Barnstaple and Lundy on 22 August 1905.

ROSEATE TERN   *Sterna dougallii*
Passage migrant

No Devon occurrence of the Roseate Tern was admitted by D & M nor was it mentioned in the *VCH*. In the *First Report* it is stated that a bird of this species was seen near Clovelly on 15 September 1928, but it is a second hand record and lacks conviction. Following this there is no further mention of it until 22 May 1949, when R. G. Adams watched two amongst a flock of mixed terns on the Exe estuary. A detailed account of this in *BB* 43 : 381 also refers to an old record of two which were said to have been seen in Plymouth Sound in 1874, but which D & M rejected. On 27 May 1951 R. G. Adams again observed two of these most beautiful birds on the Exe estuary, and on 3 May of the following year I saw one amongst a group of mixed terns in the same locality, this bird being immediately conspicuous by the rich pink of its breast.

From 1951 onwards Roseate Terns have been recorded annually on the south coast at Dawlish Warren, where the spring birds usually arrive during the first half of May, but occasionally at the end of April. The autumn passage lasts from about mid-July to late August and sometimes well into September, but varies from year to year. The numbers involved are quite small and do not often exceed a

L

maximum of ten, but fourteen were counted on 17 May 1958, eleven on 14 May 1960, thirteen on 6 May 1962, and eleven during July/ August 1965. During recent years a few have been reported from other coastal localities, including Westward Ho! and Morte Point on the north coast and Slapton and Plymouth in the south. In fact, the Plymouth records comprise two on the Hamoaze on 31 July 1965 —up to five there during July to September 1966 and up to eight during July to early September 1967 recorded by S. C. Madge. This species is very rare, however, on the north coast, and has not so far been recorded from Lundy. The nearest breeding station to Devon is the small colony in the Isles of Scilly.

The Roseate Tern has a link with Devonshire, having been first described by Colonel George Montagu in his Supplement to the *Ornithological Dictionary* from a specimen obtained in the Firth of Clyde, which was sent to him by Dr Macdougall of Glasgow.

LITTLE TERN   *Sterna albifrons*

Passage migrant

Described by D & M as 'a casual visitor of occasional occurrence in summer and autumn', the Little Tern does not appear to have been any more plentiful at that time than it is now. A regular species on both spring and autumn passage, it is recorded every year on the Exe estuary, in small parties varying from two or three to twenty or thirty birds, and it is not unusual for very small numbers to be seen there during all months from April to September.

The principal numbers recorded on the Exe include flocks of twelve in August 1936, eleven in May 1939, seventeen in September 1941 and April 1943, twenty-one in May 1947, thirty-two in May 1948, thirty-six on 21 April 1962, thirteen in April 1964 and 1965, fifteen in May 1966, fourteen in August 1966, and thirty-five on 23 April 1967. Smaller numbers of up to about ten occur irregularly at Slapton and on the Kingsbridge estuary, and most years a few are observed on the Taw and Torridge estuary, while in September 1958 a flock of about thirty terns, mostly this species, was seen in Bideford Bay.

There are no inland records, but ones and twos are occasionally reported from the other estuaries. The Little Tern is a rare bird at Lundy, where only three or four have been recorded: one by R. Perry in May 1939, one in September 1950, and one on 21 and 24 September 1955.

SANDWICH TERN   *Sterna sandvicensis*
Passage migrant

Now a common passage migrant, the Sandwich Tern was formerly scarce and irregular. D & M stated that a few occurred on the Exe in some seasons, and up to about 1930 it was still infrequent in the south and rare enough in the north to merit mention in *BB*.

Although it has been recorded annually on the Exe estuary since 1930, the maximum numbers occurring during the next decade did not exceed twelve, but in April 1941 over thirty were recorded, and sixty-three in September 1949. During the 1950s and 1960s the numbers there have mostly been over 100, with maxima of 106 in August 1952, 136 in August 1956, 160 on 9 September 1962, and 150 on 15 September 1966 and 2 August 1967.

Corresponding with the increase on the Exe, the species has become more frequent all along the south coast, where spring and autumn passage is now noted at Hope's Nose, Slapton, Prawle Point, and Wembury, among other places, with peak numbers of up to about forty in September. Since the 1950s small numbers have occurred fairly regularly on the Taw estuary, chiefly in the autumn, but occasionally in April and May, with a maximum of sixteen at Instow on 1 May 1966, in which year twenty-three were counted off Ilfracombe on 30 August.

An early migrant, the first birds usually arrive on the Exe towards the end of March, and some are present until mid-October, with the greatest numbers in the autumn; a maximum spring count of over 100 in April 1966, and a few non-breeding birds throughout most summers.

Like the other terns, it is infrequent at Lundy where singles have been noted in spring or autumn on about five occasions since 1950 and seven were present on 30 September 1959.

RAZORBILL   *Alca torda*
Resident, breeds

The Razorbill breeds on both the north and south coasts and on Lundy. Although mainly pelagic during the winter, small parties and scattered birds occur along both coasts from about August to February or March, and occasional birds enter the estuaries. A marked autumn passage is noted off Start and Prawle Points.

On the north coast 250-300 pairs breed in scattered colonies between Trentishoe and Lynton. The number does not appear to have changed to any extent at this locality during the past 20 or 30 years, but it has been variously stated as 200 pairs in 1948, 150 in 1951, 170 in 1958, over 300 in 1961, and 260 in 1967. On the south coast, however, the number of breeding pairs is very small and has decreased. The Berry Head colony contained thirty pairs in 1941 and forty to fifty in 1952, but only ten in 1962 and two in 1967. The colony at Scabbacombe Head was stated to hold ten pairs in 1948, fifty in 1949, but only five in 1962 and two or three in 1967. D & M unfortunately gave no estimates of breeding numbers or localities, but simply stated that it bred 'sparingly on some cliffs of the north and south coasts'.

The population breeding on Lundy was given by Perry as 10,500 pairs in 1939 but had decreased to only about a quarter of that number by 1953. A count of all birds present in 1962 gave a total of 2,130 individuals. No estimate was made by Loyd or previous writers, who merely spoke of 'countless' and 'immense' numbers. The colonies on Lundy are scattered all around the coast, and birds are present from about March until about mid-August.

LITTLE AUK   *Plautus alle*
Irregular winter visitor

The occurrence of the Little Auk on the coasts of Devon, and sometimes many miles inland, is almost invariably due to gales which drive them in from the Atlantic and occasionally sweep them far inland. This small seabird has been recorded in the county in 23 of the past 40 years and, of the approximately seventy-three birds reported in that period, twenty-five occurred in each of the months of December and February (including twenty-two in February 1950), nine in November, six in January, and up to four each in September, October, and March. These occurrences are fairly evenly scattered along the north and south coasts, with inland records from such places as Drewsteignton, Modbury, and Ashburton.

Although only two or three individuals normally occur in any one year, an exceptional 'wreck' occurred in mid-February 1950 when, following a period of westerly gales which drove many exhausted birds on to the west coast of Ireland and south-west England in addition to carrying others to inland counties of Britain, twenty-two were recorded in Devon. Of these, fourteen were picked up dead or exhausted in the parishes of Braunton, Georgeham, and Saunton, bordering Bideford Bay, and the remaining eight at widely separated places elsewhere in the county, both inland and on the south coast. Undoubtedly many others perished but were not found. An account of this and other smaller 'wrecks' was compiled by D. E. Sergeant and appeared in *BB* 40:122-33. A most unusual record concerns a bird in summer plumage that occurred at Dartmouth for a few days in July 1959.

The few records of this species on Lundy comprise one seen on 7 September 1926 (*BB* 25:219), the remains of one found in June 1950, and two seen off the North End on 5 November 1958.

The overall pattern of occurrences has probably not changed since the time of D & M who described the Little Auk as a 'casual visitor of occasional occurrence, generally in the autumn and winter months . . . after storms'.

GUILLEMOT *Uria aalge*

Resident, breeds

The Guillemot, like the Razorbill, breeds on the north and south coasts and on Lundy, but outnumbers the Razorbill except in north Devon. Outside the breeding season it is pelagic but small flocks and storm-driven birds occur on both coasts in all months, and an autumn movement occurs along the south coast, mainly in October.

Although they return to their breeding cliffs from about mid-March and remain until around mid-August, they are sometimes observed on the cliffs in November and December.

On the north coast it breeds on the cliffs between Heddons Mouth and Woody Bay and is apparently increasing. Whereas the population varied between twenty and fifty pairs from 1948 to 1958, it was stated to be 228 pairs in May 1967, although another account puts it at 300 individuals. D'Urban stated in the 1900s that a few pairs nested on Baggy Point, but there appear to be no subsequent records.

On the south coast 250-400 pairs breed at Berry Head and a few at Scabbacombe. At the latter they have diminished from a maximum of about sixty pairs in 1950 to two or three in 1967. The population of the Berry Head colony was given as 200-300 pairs in 1940, 500 in 1941, 300 in 1952, 750 birds in May 1961, and about 400 birds in June 1965.

The breeding population of Lundy was stated by Loyd to be immense; it was estimated by Perry to be 19,000 pairs in 1939, from which it has decreased to approximately 5,000 pairs in 1951, and 3,560 individuals in 1962, but with no apparent change since.

Birds resembling the Northern race, *U.a.aalge*, have been reported on the coasts during autumn and winter on a number of occasions.

BLACK GUILLEMOT *Cepphus grylle*

Vagrant

Of the nine occurrences of the Black Guillemot listed by D & M for the entire nineteenth century, eight refer to single birds obtained on the south coast and one to a bird picked up dead on the Taw estuary. All nine occurred during the autumn and winter and all except one were in winter plumage. D'Urban's notes for the first quarter of the present century contain no further occurrences, but six subsequent records are contained in the *Devon Reports*, which include one of a bird in complete summer plumage seen by Ernest Allen at Dawlish Warren on 4 March 1928 and reported in *BB* 21 : 303. The others refer to singles seen at Dawlish on 20 February 1938, at Prawle Point on 21 February 1939, at Start Point on 20 March 1954, a bird in summer plumage closely observed by Mrs F. E. Carter on the Taw estuary on 29 September 1956, and, lastly, one seen and adequately described by J. M. Reese at Wembury on 26 December 1962. There is no record of the Black Guillemot from Lundy.

PUFFIN *Fratercula arctica*

Resident, breeds

The Puffin, though still breeding on Lundy, has decreased tremend-ously since the 1930s, but whether the decline is due to brown rats, oil pollution, predation by Great Black-backed Gulls, or a reduction in the food supply, is not known for certain.

Perry estimated the number of breeding pairs on Lundy in 1939 as 3,500, but by 1950 it was considered by Davis to be less than 400 pairs. In 1957 the largest flock counted was 156 birds, and in 1963 was 132. By 1965 it was thought that only thirty-three pairs were left, but in 1966 the population was given as about sixty pairs.

The birds are present on Lundy from the beginning of April until the end of July, and are pelagic during the rest of the year, when individuals and storm-driven birds are not infrequently noted along the mainland coasts. The species is recorded annually off the south coast, mainly during June and July, when parties of up to twenty, usually less, and comprising adults and young, are encountered off Exmouth and elsewhere. Although occasional birds occur off both coasts during the breeding season, there are now no breeding colonies on the mainland, and the mention of one in *The Handbook* presumably refers to Berry Head where a very small colony may have existed at one time.

D & M, who did not specify any mainland colonies, stated that the Puffin occurred on some parts of the coast at all times of the year and was abundant on the north coast in summer.

PALLAS'S SANDGROUSE *Syrrhaptes paradoxus*
Vagrant

The only Devon records of this Asiatic species are those mentioned by D & M, that occurred in the great invasions of 1863 and 1888, when the birds spread over much of the British Isles. In their detailed account in *The Birds of Devon*, these authors list four occurrences involving about twenty-six birds. The largest party, consisting of thirteen birds, occurred on the sands at Slapton Ley in June 1863, and on 11 December of the same year one was shot at Heanton Court near Barnstaple. In the invasion of 1888 a flock of seven visited Lundy in May and were present for three weeks. Lastly, four or five were shot at Hartland on about 3 June 1888.

STOCK DOVE *Columba oenas*
Resident and winter visitor, breeds

The spread of this species westwards into Devon was noted by D & M, who previously regarded it as an irregular winter visitor. In their time it was scarce or absent as a breeding bird on the north coast, where it now breeds regularly not only on the cliffs but also on Braunton Burrows.

It is now widely distributed throughout the county wherever suitable conditions exist, as in parklands with old timber, sea cliffs, sand dunes, old quarries, and disused mine workings. Breeding occurs regularly along both coasts, while on Dartmoor about a dozen pairs breed at Vitifer mines, a few at Powder Mills, and others elsewhere in old timber and mine shafts.

Flocks of forty to fifty are frequently recorded during winter in all parts of the county, while larger numbers include 150 noted by R. G. Adams at Lympstone in October 1948, over 200 observed on the Avon estuary in November 1952 by D. R. Edgcombe, and 136 seen by P. J. Dare in the West Webburn valley on Dartmoor in March 1957.

On Lundy a few singles are recorded on spring or autumn passage in most years, but it does not breed.

ROCK DOVE　*Columba livia*
Formerly resident and bred

At the beginning of this century there was considerable doubt about the status of the Rock Dove in Devon. Whatever the position may then have been, the species no longer breeds and there are no satisfactory records of its occurrence during the past 40 years at least. In 1830 Dr E. Moore wrote of this species: 'I am informed . . . that this bird is found in a wild state on the south coast of Devon; and Polwhele states that it exists on the north coast, near Combe Martin, and at Lundy Island'. D'Urban in the *VCH* said that it 'appears to have been formerly much more abundant than it is at present, and is no longer found in some localities such as Lundy Island, where once it was well known'. In 1929 Loyd (*BSED*) believed that a few pairs of genuinely wild birds still bred at Beer Head, and in 1932 Dr Elliston Wright listed it as occurring near Braunton.

In the *Devon Report* for 1929-30 it is stated that a pair of Rock Doves were positively identified at Bolt Tail and were probably nesting, while *BB* 25:218 records that a bird of this species was said to have been identified on Lundy on 31 August 1927. The few records contained in the early *Devon* and *Lundy Reports* must be regarded as very doubtful and most probably refer to feral pigeons.

WOOD PIGEON *Columba palumbus*
Resident and winter visitor, breeds

Resident and abundant, the Wood Pigeon breeds very commonly throughout the county, from the city parks to the highest plantations on Dartmoor and in comparatively isolated copses such as Wistmans Wood. It breeds abundantly in all the state forests. According to D & M it was already becoming a serious pest to agriculture in their time, and has continued to increase since.

The *Devon Reports* contain many records of winter flocks of up to 1,000 birds, and up to 2,000 were reported in the fields above the cliffs at Chiselbury Bay in December 1963. Among the many records of autumn migratory movements, both inland and coastal, the following may be mentioned: 2,000 flying south-west along the coast at Exmouth on 16 November 1947; about 5,000 in November 1952 reported by R. G. Adams in flocks of up to 600 flying westwards over Lympstone; and again, about 4,000 passing westwards in one hour on 5 November 1959. H. G. Hurrell has recorded flocks moving south and south-west over Wrangaton during October and November.

The one to four pairs which breed regularly on Lundy leave the island in October and return during February or March. Spring and autumn passage movements, involving peak numbers of twenty or more, are recorded in most years.

TURTLE DOVE *Streptopelia turtur*
Summer visitor, breeds

The Turtle Dove, a summer resident, is locally distributed, being confined mainly to the country stretching eastwards from Dartmoor to the Dorset border, and northwards from Dartmoor to the Bideford/Barnstaple area and the Somerset border. Although it may breed sporadically in the west and south-west, it is unusual in these parts and doubtless most of the records refer to migrants.

It has always bred regularly at Braunton Burrows and Saunton, with four or five pairs in 1960 and frequent records of around a dozen birds during July and August. There are definite nesting records from Brendon, Umberleigh, Drewsteignton, and Sandford, and the species seems to be well established in the Clyst St Mary/East Budleigh area, where flocks of up to thirty have been recorded during June and July in a number of years since the 1940s. Except for Dartmoor, the west and south-west areas, it is presumed to breed in many other localities and birds are fairly regularly reported during May and June, though the numbers vary from year to year. The Turtle Dove is rare on Dartmoor itself, where there are occasional records, mostly of singles. In 1967, however, one was reported singing at Fernworthy during July.

Although there is no reliable breeding record for Lundy, the species occurs annually on spring passage from early May almost to the end of June, followed by a small return passage during July/September. A larger number than usual was recorded during the second half of May 1964, when flocks of up to thirty occurred on several days.

COLLARED DOVE   *Streptopelia decaocto*

Resident, breeds

The rapidly increasing Collared Dove, which colonised Britain in 1955, was originally regarded as having been first recorded in Devon in 1961, when one or two birds were observed on Lundy at the end of May. In 1963, however, it was reported that the species first appeared in the centre of Plymouth in 1960, and by 1963 had increased to thirty.

In September 1961 four appeared at Budleigh Salterton and three at Instow, where they bred in 1962, in addition to Kingsbridge, Plymouth, and Pilton. By 1963 the numbers had increased to at least twenty-one at Budleigh Salterton, eight at Kingsbridge, eighteen at Instow, sixteen at Pilton, thirty at Plymouth, and twelve at Knowle, with first records from Woodbury, Exmouth, Lympstone, Bideford, and Fremington. In 1964 sixty-two were counted at Budleigh Salterton, and it spread to Dawlish, Zeal Monachorum, Torrington, Beaford, and Braunton, as well as to fresh localities in Plymouth, in two of which a total of eighty-four birds were counted in the autumn of 1965.

The total at Budleigh Salterton was thought to exceed 200 in 1966 and the birds were increasing and spreading into other towns and villages, including Postbridge where two were seen in June. By the end of 1967 it was established in most if not all the main towns and a great many of the villages.

Meanwhile, ones and twos were again recorded on Lundy from 1963 onwards, increasing to thirteen on 1 May 1966, of which one pair remained throughout the summer but was not known to have bred.

CUCKOO   *Cuculus canorus*
Summer visitor, breeds

The Cuckoo, a summer migrant arriving on average in mid-April and leaving in July to September, is widely distributed throughout the county, occurring in all areas. It is equally at home on the lowland coastal marshes as on the highest moorland of Dartmoor and Exmoor.

Although stated in the *VCH* to be an abundant summer migrant, and described as common in most subsequent local lists, there is no doubt that it has decreased in numbers during the past 10-20 years. The *Devon Reports* for 1964 and 1965 both remarked on its being less plentiful, though it was stated to be rather more numerous in 1966. Dare and Hamilton state that although common on the open moorland of Dartmoor it is generally considered to have decreased in the last 20 years.

On Lundy it occurs regularly in small numbers on spring and autumn migration, the number in any one day rarely exceeding three or four, but eight were recorded on 1 August 1948 and seven on 19 May 1962. Breeding by one or two pairs has been suspected in a number of years including 1939, 1949, probably 1950, definitely 1952, and probably 1955. There are no breeding records since 1955. It is stated by Loyd to have bred on Lundy during the 1920s.

YELLOW-BILLED CUCKOO   *Coccyzus americanus*
Vagrant

The inclusion of this trans-Atlantic drift-migrant rests on the single record of a specimen which was picked up dead beneath the old lighthouse on Lundy some time in October 1874. D & M relate that a detailed description of it was given to them by the Rev H. G. Heaven, who at that time owned and lived on the island. It is of interest to note that during the same month another example of this species occurred at Hainault in Belgium, while an American Robin was reported at Heligoland; all three birds had probably been carried across the Atlantic in the same airstream. The occurrence of this non-parasitic American cuckoo was listed in square brackets by D'Urban in the *VCH* but the record was accepted by *The Handbook*.

BLACK-BILLED CUCKOO   *Coccyzus erythrophthalmus*
Vagrant

As in the case of the Yellow-billed, the Black-billed Cuckoo has been recorded only once in the county, and that also was on Lundy, where one was reported on 17 October 1967. An even rarer bird than the very similar, preceding species, it has been recorded in the British Isles on only about five previous occasions.

BB 61 : 346 adds that this example, a first year female, was found dead on 20 October, and the specimen is now in Leicester Museum.

BARN OWL   *Tyto alba*
Resident, breeds

The Barn Owl, according to D & M, was still common and generally distributed at the close of last century, but was subject to much senseless persecution. A resident, sedentary spcies, breeding in all parts of the county except high moorland, it is still widely though thinly distributed, but despite the cessation of most persecution, it seems barely able to hold its own.

Good numbers were recorded during the 1930s and it was several times reported to be increasing, but it suffered a setback in the severe winter of 1947 and possibly again in 1963. Summarising an enquiry into its status, in 1960-64, P. F. Goodfellow considered it to be widely spread in agricultural areas mainly below 500 ft, but nowhere common. The population of the district between Axminster, Colyford, and Seaton was stated by R. Cottrill to be at least thirty-one birds in 1962.

Around Postbridge it was reported to be resident at several farms up to 1947, when it was wiped out. It has since bred in this area in 1962 and 1966 and possibly at a second site in 1965. Barn Owls, not necessarily breeding pairs, were reported in 1967 from fifty-five localities, covering all parts of Devon.

For Lundy, P. Davis quotes six dated occurrences up to 1939, including a pair present for over a year in 1922-3, but which were not proved to nest. The only subsequent record is of one observed on 10 October 1957.

The Dark-breasted race (*T.a. guttata*) from Europe has been recorded on three occasions: at Ottery St Mary on 8 October 1944, *BB* 38:175; Teign estuary 24 January 1948, by G. H. Gush; and near Newton Ferrers by O. D. Hunt on 25 September 1949, *BB* 43:225.

SCOPS OWL   *Otus scops*

Vagrant

D & M knew of no Devon record of the Scops Owl. The first and only occurrence in the county of this very small owl, so typical of the Mediterranean countries, is of a single bird which was carefully observed at Slapton on 8 May 1955 by M. R. Edmonds, whose detailed description is given on p 36 of the *Devon Report* for 1955.

SNOWY OWL   *Nyctea scandiaca*

Vagrant

Although occurring regularly as a winter visitor in the extreme north of Britain, where in fact it bred in 1967, the Snowy Owl rarely wanders so far south as Devon, though an exceptional visitation occurred in the early months of 1945 when these arctic birds were recorded in Somerset, Devon, and Cornwall, as well as elsewhere. It has been recorded under a dozen times in the county, and some of these records are doubtful or not fully substantiated. D & M give particulars of four occurrences: the first said to have been obtained at Exmouth some time prior to 1851; one killed near Plymouth in December 1838; one shot at Ditsworthy, Dartmoor, on 13 March 1876; and another was trapped on Exmoor about ten days later. The actual locality of the last is not stated and it may well have been in Somerset.

The *Devon Reports* contain details of four records : a bird probably of this species seen at Huntsham on 26 December 1932; one which arrived at Plymouth on a ship on 24 October 1932, having alighted on the vessel at sea; a bird seen by several observers in the vicinity of Emmetts Grange, Exmoor, about half a mile over the Somerset border, between 26 January and 1 February 1945; and one watched by B. G. Lampard-Vachell on Torrington Common on 16 February 1947. Of a number of Snowy Owls seen in different parts of England during the early part of 1945, it is reported in *BB* 38 : 374-5 that one was seen near Cullompton during April and another near Hatherleigh on 12 April. There is no record of this species for Lundy.

LITTLE OWL   *Athene noctua*
Resident, breeds

Of the seven occurrences of the Little Owl listed by D & M, most were thought to refer to escaped birds, though some may have been genuine vagrants. The species was introduced into eastern England in the late nineteenth century. The first record for Devon for the present century was one seen at Thurlestone in November 1911 and reported in *BB* 5 : 333. D'Urban's MS mentions three other occurrences in 1916, 1917, and 1919, and states that in 1922 it was established at Budleigh Salterton.

Loyd records that from about 1919 it increased rapidly in east Devon and in 1923 forty were killed in the parish of Branscombe alone. It was first reported in north Devon in 1922—*BB* 16 : 309. The *Devon Report* for 1930-31 stated it to be established in most parts of the county, but not in any great numbers.

The Little Owl continued to increase until around the mid-1940s when it was reduced in places by the severe winter of 1946-7 and later by that of 1962-3. In 1961 it was considered to be decreasing, although recorded from nine localities in the Exmouth/Woodbury area. Summarising its status in 1964, P. F. Goodfellow described it as a very local bird, nowhere common, and having decreased during the past decade. In the *Devon Report* for 1966 it was recorded from over thirty sites, mostly in the southern half of the county.

Although formerly breeding in the Postbridge area of Dartmoor, it was wiped out in 1947 and has since occurred only casually.

P. Davis lists records for Lundy in June 1933 and November 1944; a further single was reported there on 1 June 1955.

TAWNY OWL   *Strix aluco*
Resident, breeds

According to D & M the Tawny Owl was resident and generally distributed in wooded districts, but was not so plentiful as formerly. The decrease, which they attributed to persecution, has long since been arrested, and this species is now by far the commonest owl, with no apparent change of status during the past 30-40 years. It is very widely distributed and common throughout the county, and occurs over much of Dartmoor, particularly in the fir plantations and in the vicinity of farms. P. J. Dare considered there were ten or twelve pairs in the 16 square miles around Postbridge, and in 1957 reported nine breeding pairs in an area of only about 6 square miles. The Tawny Owl also occurs commonly in towns and cities, where it is resident in many parks. It also makes frequent use of nesting boxes provided in suitably timbered gardens.

On Lundy there was no definite record of it until 1957, when one was seen on 14 October. The only other record for the island was in the following year, when another single bird occurred on 8 April.

LONG-EARED OWL   *Asio otus*
Irregular visitor, may have bred

Its essentially nocturnal habits and preference for woodlands, especially coniferous woods, combine to make the Long-eared Owl a particularly unobtrusive species whose status is consequently difficult to define with accuracy.

According to D & M it was numerous in the winter of 1873-4, in the autumn of 1879, and January 1886, but otherwise occurred somewhere in the county almost annually. A pair were said to have nested on Haldon in 1863, and D & M said they once found a nest in north Devon, but gave no date.

The reported occurrences for the present century are very much fewer and number only about twenty-two, including five for Lundy. They comprise all singles, except for a pair on one occasion, and are as follows : near Haytor 1 August 1942, West Down near Ilfracombe 22 November 1942, Huntshaw Woods 19 April and Weare Giffard 22 September 1946, near Buckland Brewer 12 October and 13 December 1947, Dalwood Hill near Axminster (one observed many times during 1949 and two on one occasion), Huntshaw Wood 26 January and Landcross 26 March 1955, Tavistock 3 February 1956, Arlington 7 June 1957, Lynton 5 October 1961, near Newton Abbot 5 January and Slapton 7 September 1963, Plymbridge 29 May 1964, and Prawle Point 16 April 1966.

The *Ilfracombe Fauna & Flora* quotes Dr Elliston Wright as saying that he last found a nest at Braunton in 1931. G. H. Gush, who recorded several of the north Devon occurrences, believed that this owl bred in the Torridge valley woodlands, but was unable to obtain proof; and in the early 1940s B. G. Lampard-Vachell was informed of a nest at Huntshaw, said to have been of this species.

Of the five records for Lundy, the first relates to a bird caught in a rabbit snare in the winter of 1929-30 (*BB* 25 : 216), one was trapped and ringed on 15 October 1954, another was present from 25 to 27 May 1962, one was trapped on 28 July 1966, and the fifth was seen on 20 March 1967.

SHORT-EARED OWL *Asio flammeus*

Winter visitor and passage migrant, has bred

Although it is known to have bred on several occasions during the present century, the Short-eared Owl is principally a winter visitor and passage migrant, usually in very small numbers. It has been recorded in all but about five of the years since 1931.

A bird of open, marshy country, moorland, and sand dunes, it occurs mainly from October to March. The most favoured localities are the sand dunes and marshes of the Taw and Exe estuaries, where it occurs almost annually, but it has been recorded in many coastal areas, including Slapton and Start Point in the south, and Hartland, Baggy, and Morte Points in the north. Inland occurrences are all from moorland areas, including Beaford Moor, Molland Common, west Exmoor, and Tamar Lake; there are several records from Dartmoor

Page 189: *Cirl Bunting, male feeding young. A local species with a mainly coastal distribution, the Cirl Bunting is adversely affected by severe winters*

Page 190: Woodbury Common. Ornithologically a most interesting area, it unfortunately suffers considerable disturbance from military training, quarrying and picnickers

including Rippon Tor, Yealm Head, Bellever, Cadover Bridge, and Brockhill Ford.

The most recorded was twelve at Lynton Cross in December 1942, and six on Molland Common in January 1967, but it usually occurs singly.

The only record of breeding known to D & M was near Braunton in 1893. The *Zoologist* 1907 records that a pair bred on Braunton Burrows in 1906 and probably annually about that time. D'Urban's MS adds 1903 and 1904 for this locality, where it is also reported to have nested in 1931. Dr F. R. Elliston Wright reported that three pairs nested near Braunton in 1943.

On Lundy this species has been recorded annually since 1950, mostly in April/May and September to November, usually singly, but four on 16 November 1958 and 14 March 1959.

NIGHTJAR *Caprimulgus europaeus*
Summer visitor, breeds

Although the Nightjar is still thinly and locally distributed as a summer resident, on suitable commons, bracken-clad hillsides, moorlands, and woods, there is no doubt at all that it has decreased during the past 20 years. Loss of habitat and increased disturbance, whilst accounting for part of the decrease, may not be the main causes.

It is difficult, however, to define accurately its present status, because of its crepuscular habits, the ease with which it can be overlooked, and the great extent of apparently suitable habitat.

In the *VCH* it was stated to be 'very numerous in suitable localities throughout the county, especially on the borders of Dartmoor'. Breeding and presumed breeding localities occupied during the 1960s include Maiden Down and other localities in the Blackdown Hills, Budleigh Salterton, East Budleigh and Woodbury Commons, Haldon, Chudleigh Heath, Sandygate, Kingswear, Braunton Burrows, Buckland Filleigh, Highampton, Yelverton, Denham Bridge, and Bickleigh Vale. Although it was stated in *Bird Study* 9: 105 to be no longer breeding on Dartmoor, five pairs were reported at Soussons Plantation in July 1965, and other birds were noted at Bellever, Vitifer mines, and Wistmans Wood during the 1960s. Ten pairs were reported at Plym Forest in 1964, but it has certainly decreased since, while at Haldon there were at least seven resident pairs in 1966.

M

On Lundy ones and twos have been recorded in fifteen of the years 1947-66, mostly in August/September, less frequently in May, and occasionally in June and October. Breeding was suspected but not proved in 1942-3.

SWIFT   *Apus apus*

Summer visitor, breeds

An abundant summer visitor which stays for only the months of May, June, and July, the Swift is widely distributed throughout the county and breeds in every village, town, and city, wherever there are suitable old buildings, thatched cottages, and church towers. On high moorland it is restricted by lack of breeding sites, but it occurs commonly on feeding flights over Dartmoor and breeds at 1,350 ft at Princetown.

Although it has often been said to breed in the coastal cliffs, and Loyd stated that it still did so at Beer Head, while others have reported cliff nesting on Lundy and elsewhere, there appears to be no evidence in support of this, nor is there any definite record of breeding at all on Lundy.

One of the last species to arrive and the first to leave, the Swift does not normally appear until the end of April and beginning of May, and most have gone by the first week of August, although autumn passage continues during that month. What was almost certainly one of the characteristic weather movements of this species involved a flock of possibly 2,000 which were seen drifting over Lympstone on 10 June 1942.

A regular and common passage migrant on Lundy, most of the birds occur from May to August, with early arrivals in April and departing stragglers in September. The maximum number recorded at one time over the island is about 400 on 17 August 1958, but the numbers fluctuate considerably and the peak is usually much less than this.

ALPINE SWIFT   *Apus melba*

Vagrant

The Alpine Swift, a straggler from southern Europe, has been recorded in the county on a number of occasions. D & M relate that an immature bird was shot, and two others seen, at Ilfracombe on 4 October 1876, in company with a flock of the common species. The *Field* for 22 September 1906 reported a pair which were said to have been seen at Withycombe Raleigh, Exmouth, in August of that year. A supposed example of this species was recorded in *BB* 2:140 as having been seen at Lynmouth, but the editors were somewhat sceptical of this undated record. The same journal (18:114) notes that one of these large swifts was shot at Start Point on 14 April 1924, while a subsequent issue (23:306) states that one was picked up dead at Newton Ferrers on 11 March 1930.

The first record in the *Devon Reports* of this easily identified swift concerns a single bird which was observed at Lympstone on 2 October 1938 by R. G. Adams who had an excellent view of it for several minutes. On 15 August 1959 another single bird was identified by B. S. Meadows amongst a flock of Swifts over the Taw estuary, being immediately recognised by its much larger size and mainly white underparts. The latest accepted record refers to a single bird seen at Berry Head by B. C. Cave on 7 October 1967, *BB* 61:347.

In 1959 the Alpine Swift was recorded for the first time on Lundy where one was present from 9 to 11 May and was watched by a number of islanders, sometimes at very close range. There have been two subsequent records for Lundy, one on 25 April 1962 and another on 28 September 1965.

KINGFISHER   *Alcedo atthis*

Resident, breeds

Although far from plentiful, the Kingfisher is widely distributed along most streams and rivers, except the fast-flowing upper reaches of the moorland rivers. After the breeding season most birds move downstream in late July and August, and from then until March they frequent the estuaries and to a lesser extent the coast, though in some localities they breed beside tidal water.

It has not been recorded in the Postbridge district and there appear to be no records for other parts of central Dartmoor, though it occurs up to the edge of the moor, as at Double Waters on the Tavy, and on western Exmoor it has been recorded on the East Lyn at Brendon.

The *Devon Reports* for 1960-62 mention 'many breeding records', 'widely distributed' and 'many known breeding areas' but the very severe winter of 1962-3 almost exterminated it, and although by 1967 the numbers had not fully recovered, records from all the main rivers throughout the county indicate that it is regaining its former status.

Mist-netting at Slapton has revealed a small coastal movement between July and September, mostly of juveniles, with twenty-five ringed 1961-2, none in 1963, and twenty-two 1964-7. Further evidence of movement is shown by the occasional records of singles on Lundy where seven or so have been seen since 1947, with one dated record for March and the remainder from July to September.

Except for fluctuations due to hard winters, the status of the Kingfisher does not appear to have changed appreciably since the early 1900s when D'Urban described it as a common resident in the greater part of the county and breeding in many localities.

BEE-EATER *Merops apiaster*

Vagrant

The occurrences of the brilliantly coloured Bee-eater in Devon are few and far between. Of the six or seven listed by D & M as having been obtained or seen during the nineteenth century, half are inadequately authenticated, and all refer to single birds in the south of the county.

A vagrant from southern Europe, this graceful and unmistakable bird is highly gregarious and tends to occur in small parties rather than singly. Of the only two fully substantiated records for Devon during the present century, the first relates to a flock of six which appeared on Lundy on 19 May 1940 and remained on the island until 26 or 27 May, returning again for a brief visit between 1 and 3 June. This occurrence was reported in BB 34:163 by F. W. Gade, but the birds were also seen by H. A. Jackman who was staying on the island at the time. The second refers to a single bird seen at Ottery St Mary on 20-21 October 1963 which was fully described.

A very probable occurrence, contained in the *Devon Report* for 1949, relates to a party of four which frequented a spot between Beer and Seaton for about three weeks during June 1949 and were correctly described by an R.A.C. patrolman who reported them to the Rev F. C. Butters.

ROLLER   *Coracias garrulus*
Vagrant

Another species in the long list of vagrants to the British Isles is the exotic Roller, a summer resident in southern and eastern Europe, which has been reliably recorded in Devon on nine occasions, all between the months of April to October. The accepted records, including four listed by D & M for last century, are as follows:

1841	Dalditch, Budleigh Salterton, one obtained in September
c1850	Barnstaple, one killed (no date)
1866	Near Yealmpton, one shot, 21 June
1866	Alphington, Exeter, one shot, 20 October
1911	Near Chagford, one present for several days from 3 August, *BB* 5:136
1923	Near Budleigh Salterton, one seen by W. Walmesley White, 11 April, *BB* 17:86
1933	Kingsbridge, one reported seen in July
1949	Lundy, one, probably a female, seen by E. H. Ware, 25 August
1956	Exeter, one closely observed by F. R. Smith in his garden, 28 May

The *Devon Report* for 1939 contains a somewhat doubtful record of a bird thought to be a Roller, seen near Bovey Tracey on the unusual date of 18 February. It will be seen from these records that most of the occurrences have been on or near the south coast, but there is one from north Devon, one from Dartmoor, and one from Lundy.

HOOPOE   *Upupa epops*
Passage migrant

As long ago as 1830, in one of the earliest lists of the birds of Devon published, Dr E. Moore wrote of the Hoopoe: 'This beautiful bird is not infrequently met with in Devonshire'. Some sixty years later D & M summarised its status in the county as: 'A summer migrant of irregular occurrence, principally in spring and autumn, most frequently obtained in the South Hams, especially in the Plymouth and Kingsbridge districts, and on Lundy Island. It has been known to breed in Devon. A nest with four young was taken in a wood close to the house at Morwell in the parish of Tavistock'. This is followed by a list of many birds shot in different parts of the county during the nineteenth century.

The Devon Reports indicate that the Hoopoe has been recorded in Devon in all but 4 of the years from 1931 to 1967 inclusive, the exceptions being 1936, 1938, 1939, and 1940. Of some 240 birds recorded during this period, including about forty from Lundy during the past 20 years, over three-quarters refer to the spring passage, March to June, and just under a quarter to autumn passage, July to October. Of this total, over ninety occurred during April, fifty-five in May, about twenty-five each in March and September, a dozen each in June, August, and October, and two odd birds in July. The occurrences are spread over practically the whole of the county, with quite a number of birds seen on different parts of Dartmoor, including Fernworthy, Peter Tavy, Two Bridges, and Bellever, with no less than four at the last named, during late April and early May 1964. Although a number have been reported from localities in north Devon, far more have been seen along the south coast than elsewhere.

Whether because of the greater number of bird-watchers in recent years, or because the Hoopoe is becoming more frequent, far more have been seen during the past 20 years than in the previous 20; in fact, about 210 of the approximately 240 mentioned refer to the 21 years from 1947 to 1967. As many as fourteen were reported in 1956, twenty-four in 1958, fourteen in 1960, and sixteen in 1964, including the Lundy records. As already stated, about forty have been recorded there since the establishment of the Bird Observatory. It has for long been considered regular on Lundy, but prior to 1947 there was little systematic observation, and the records are fragmentary.

The great majority of the county records are of single birds, very occasionally two, and only exceptionally three or more. Three or four were seen on Lundy on 16 April 1951.

GREEN WOODPECKER *Picus viridis*

Resident, breeds

A resident and sedentary species, the Green Woodpecker is evenly distributed in singles and pairs throughout the entire county wherever there are mature deciduous trees. Although normally absent from treeless moorland, it breeds in all suitable areas on Dartmoor, where it has been recorded at approximately 1,400 ft at Princetown and was noted as breeding at 1,300 ft in the Postbridge district in 1958. It occasionally occurs far out on the moor, away from trees, having been observed at Avon Head in July 1942 and near Cranmere Pool in August 1963.

This woodpecker also occurs quite frequently on the coast, where it has been observed feeding on both the cliff face and the beach. The numbers were reduced slightly by the severe winter of 1962-3, especially on Dartmoor, but except for fluctuations due to weather, there is no evidence of any long term change of status. It is non-migratory and has not been recorded on Lundy.

GREAT SPOTTED WOODPECKER   *Dendrocopos major*
Resident, breeds

Judged by the old records, the Great Spotted Woodpecker has increased quite considerably during the present century. In 1830 Dr E. Moore described it as 'rather scarce' whereas he regarded the Green Woodpecker as 'common all the year'. Pidsley in 1891 spoke of it as 'a scarce resident in our more extensive woodlands, rare as a breeding species', and D & M considered it 'resident but more often met with in winter and spring than at other times of the year. . . . Though not common anywhere, this woodpecker appears to be generally distributed'.

At the present time the species is widely distributed and breeds in all suitable woodlands throughout the county, except for the higher parts of central Dartmoor. Although P. J. Dare recorded two or three pairs in the Postbridge district, they were seen chiefly in the West Webburn valley, and the species is absent from the fir plantations at Bellever and Soussons. L. A. Harvey, in *Dartmoor*, mentions a nest at 1,250 ft but confirms that the species is scarce or absent from the moor. It occurs, however, in the heavily wooded river valleys where the rivers leave the moor. Since the late 1950s the species has become a regular visitor to bird-tables in many parts of the county, and is acquiring the habit of bringing its young to feed there also.

The Great Spotted Woodpecker was able to withstand the arctic conditions of the 1962-3 winter better than many resident species, and was well recorded during 1963, while by 1964 it was reported to be in 'good numbers everywhere'. To what extent the resident numbers are increased by winter visitors is difficult to say, but that some migratory movement does occur is proved by the autumnal visits of the species to Lundy. Although entirely absent as a breeding species, seven individuals have been recorded there in six years since 1949, singles having been seen in the months from August to November (usually September), in 1949, 1957, 1959, 1962, 1964, and 1965. Of two birds observed during October and November 1962, both were caught and ringed, and proved to be first winter birds. A single bird during early October 1949, although not examined in the hand, was thought to be of the Northern race, *D.m. major*.

LESSER SPOTTED WOODPECKER *Dendrocopos minor*
Resident, breeds

The least common of the three species, the Lesser Spotted Woodpecker is well but thinly distributed throughout the county, except on high ground, but its unobtrusiveness tends to make it appear scarcer than it really it. It occurs in deciduous woodland, parkland, orchards, and gardens with suitable timber.

Since 1950 breeding has been recorded in all parts of Devon, including the west and north-west, but not Exmoor (although singles have been observed at Wistlandpound) and not central Dartmoor. At Yarner, however, on the eastern fringe, it breeds annually, and it has been recorded at a number of other fringe areas; but it is not recorded in the Postbridge area.

A temporary drop in the population, caused by the extremely hard winter of 1962-3, has probably been made good by now, and there is nothing to suggest that its status has changed since D & M's time. A sedentary and non-migratory species, it has not been recorded on Lundy.

WRYNECK *Jynx torquilla*
Scarce passage migrant, has bred

D & M regarded the Wryneck as a rare bird, occurring chiefly in spring and autumn, but they knew of no definite record of breeding.

According to Loyd in *BB* 19:132, a pair nested at Branscombe in 1925 and probably in 1924. Although W. Walmesley White in *Bird Life in Devon* stated that it was spreading westwards into Devon and that he occasionally recorded it around Budleigh Salterton during the 1920s, it now seems probable that most of the occurrences related to spring passage birds, as there is no record of its breeding. A pair are believed to have bred at Crapstone, Yelverton in 1954.

The *Devon Reports* contain about thirty records for the mainland during the past 40 years, mostly in coastal areas, including a few from the north, but also a few inland. The majority occurred during the months of April (twelve) and September (six), with one or two in all other months from March to October.

Since 1947 almost twenty have been reported on Lundy, all on spring and autumn passage, mainly in April and September, but not annually.

Although the Wryneck has most probably decreased in Devon, the number of records is not appreciably fewer in recent years than formerly.

BIMACULATED LARK   *Melanocorypha bimaculata*
Vagrant

The first and only known occurrence in Britain of this Asiatic species was recorded on Lundy by Michael Jones, who first saw the bird on 7 May 1962. It was present in the same area of short pasture for four further days, until 11 May, during which time it was watched for several hours by Michael Jones, the Warden and his assistant, Richard Carden, who both took copious notes of the bird, thinking it to be a Calandra Lark. From their detailed description and field sketches, the Rarities Committee had no hesitation in identifying it as a Bimaculated Lark, a species new to Britain. Full accounts of this most interesting occurrence are given in *BB* 58:309-12 and the *Lundy Report* for 1963-4 pp 14-15.

SHORT-TOED LARK   *Calandrella cinerea*
Vagrant

Devon can claim only one record of this Mediterranean species, a single example of which was identified by W. Walmesley White amongst a small flock of Skylarks and Meadow Pipits on Dawlish Warren on 2 August 1928. A description of this bird is given by him in *BB* 22 : 108 and is included in *The Handbook*. D & M refer, in brackets, to a bird they observed on Braunton Burrows which they thought was this species but were unable to obtain.

CRESTED LARK   *Galerida cristata*

Vagrant

Although a widespread and common species on the Continent, the Crested Lark is an extremely sedentary bird which is rarely recorded in Britain. Until 1958 its claim for inclusion in the Devon list rested on two somewhat doubtful records which, although accepted by D & M, were not sufficiently well authenticated for inclusion in *The Handbook*. The first of these refers to a bird shot on Braunton Burrows about 1851, and the second to a pair which Lord Lilford informed D'Urban he had observed at Slapton Ley during July 1852. At the end of 1958, however, a single bird was identified on some waste ground beside the estuary at Exmouth by W. H. Tucker, who noted the characteristic crest, the long, curved bill, and buff edges to the short tail. This bird, which allowed of a very close approach, remained in the same area from 29 December, when it was first seen, until 3 January 1959, during which time it was observed by a number of bird watchers. This record was accepted by the Rarities Committee, *BB* 53 : 167.

WOODLARK   *Lullula arborea*

Resident, breeds

Described by D & M as resident and common in some districts, this delightful bird has greatly decreased during the past 20 years and was further severely reduced by the hard winter of 1963.

An enquiry into the status of the Woodlark in 1957-9 showed that although it was still thinly distributed in suitable, hilly country over a great part of Devon, including all the fringes of Dartmoor, it was much less plentiful than during the 1940s. In 1949, for instance, three winter flocks of thirty, eighteen, and forty were reported, and in January 1952 one of fifty.

Although in 1966 there were records from twenty localities in all parts of the county, including a winter flock of twenty, there was no indication of any recovery in numbers. The *Devon Report* for 1967 recorded about twenty singing males, including five at Dartington.

The Woodlark occurs irregularly on Lundy, where just over twenty individuals, in ones and twos, were recorded between 1951 and 1960, but only one since, in October 1966. Over half occurred from September to November, the remainder from January to May.

SKYLARK   *Alauda arvensis*

Resident and winter visitor, breeds

An abundant breeding bird throughout Devon, the Skylark is widely distributed in all open country. It breeds commonly over the whole of Dartmoor and Exmoor, but usually deserts high moors in mid-winter, when foraging flocks of 100-300 resort to lower agricultural land.

Westerly movements along the coast in autumn are noted in some but not all years, as are irregular hard-weather movements, sometimes involving thousands of birds. In 1964, for example, 5,000 were observed moving south-south-west over the Axe estuary on 13 January, where on 28 December about 20,000 were seen flying west or south-west. On 1 November of the same year over 600 were noted coming in from the sea at Wembury.

The Skylark breeds annually on Lundy, where the population was recorded as thirty pairs in 1939, around fifteen during the 1950s, fifty pairs in 1962, and forty in 1966. The winter population is very small; there is often a marked autumn passage with peaks of up to 300, usually less, a small spring passage, and irregular weather movements.

There is nothing to suggest any change of status in Devon since the last century.

SHORE LARK   *Eremophila alpestris*

Vagrant

Devon, lying well to the west of the somewhat restricted wintering area of this northern species in England, receives only the occasional bird or small flock. In fact, D & M were able to cite only one authentic occurrence of the Shore Lark, namely that of a flock seen on Northam Burrows on 2 January 1875, one bird of which was obtained. Two other supposed occurrences, at Paignton and Dawlish, were rejected by them as being unsatisfactory.

A bird of the seashore and coastal areas during the winter, the Shore Lark rarely occurs inland, the only such record for Devon referring to a bird in a very bedraggled condition seen at Lifton during October 1928, having probably been carried there by a gale. The seven reported occurrences for the present century are all between the months of October to March, and comprise single birds or small parties of up to five. The first of these relates to a female which frequented the Otter estuary from 7 December 1915 to 18 January 1916 and was reported by W. Walmesley White in *BB* 9:276 and the *Devon Report* for 1928-9. The latter journal also includes the Lifton record already mentioned. The next record refers to a single bird reported seen on Dawlish Warren on 11 October 1941, but is a secondhand record and lacks evidence. In 1947 a party of five Shore Larks was seen by G. H. Gush at Fremington Marsh on the Taw estuary on 3 November and recorded in the *Devon Report* for the following year, 1948. A single male was seen by a number of birdwatchers at Dawlish Warren between 2 and 7 November 1954, and a male at Wrafton on the Taw estuary by G. H. Gush on 16 February 1955, both being listed in the *Devon Reports* for the years concerned. The most recent occurrence is that of a party of four, accompanied at times by Snow Buntings, observed by W. H. Tucker on Northam Burrows on 5 November 1963.

The only record of the Shore Lark on Lundy, apart from Chanter's list of doubtful occurrences, is that in *BL 66* referring to a single bird which was present for several days from 24 March 1944 and was reported by F. W. Gade. This record, incidentally, is given in error as 24 April in the *Ilfracombe Fauna & Flora*.

SWALLOW   *Hirundo rustica*

Summer visitor and passage migrant, breeds

A common summer visitor and abundant passage migrant, the Swallow is well distributed throughout the county, nesting mainly in farm buildings but many other buildings, often near water, and also in isolated huts in exposed positions on Dartmoor.

The first arrivals frequently occur at the end of March, with the main influx during April, and passage continuing well into May. In spring there is much overland migration, with small flocks moving northwards up the river valleys, in addition to a well defined route, generally north-eastwards, along the north coast of Cornwall and Devon, and also northwards over Lundy.

Before their departure around mid-September, great masses congregate and roost in the reed beds at Slapton, where the peak numbers have been estimated at 10,000 in several years during the 1960s. Similarly, up to 10,000 roost on the Axe estuary during September. One ringed at Topsham reed bed on 14 September 1965 was retrapped at Rosherville Dam, Johannesburg, on 19 December 1965. Occasional late birds are seen in November.

On Lundy sporadic breeding by one pair occurred in 1952, 1954, 1959, 1962, and 1963, in addition to a number of previous years listed by P. Davis. Great numbers of Swallows pass through the island on spring and autumn passage, with the main movements in April/May and September/October. A spring peak of 665 was recorded on 16 May 1955 and the highest autumn peak was 2,500 on 8 September 1960.

There are insufficient numerical records to indicate any significant change of status since D & M's time.

RED-RUMPED SWALLOW   *Hirundo daurica*

Vagrant

The Red-rumped Swallow, a very rare vagrant to the British Isles, has been reliably recorded once in Devon, while another record which was square bracketed is almost certainly correct. A single bird, in company with a Swallow and two Sand Martins, was seen on the west side of Lundy by Peter Davis and John Ogilvie on 27 March 1952. It was present throughout the day and was observed at distances of less than 10 ft. Full details of the occurrence are given in *BB* 46 : 264. It is of interest to note that this species also occurred in Norfolk on 6 March and in Ireland on 10 April of the same year.

An earlier record relating to two birds seen at Sidmouth on 25

April 1947, in company with Swallows, and House and Sand Martins, was included in brackets in the *Devon Report* for 1947 because the editorial committee, although believing it to be a good record, considered it safer to do this in view of the extreme rarity of the species in Britain. The descriptions of the birds, however, by two independent observers, leave no reason at all to doubt the accuracy of the record.

Since writing this account, one was observed at Turf on the Exe estuary on 21 April 1968.

HOUSE MARTIN *Delichon urbica*

Summer visitor, breeds

A common and widely distributed summer resident, the House Martin breeds throughout the county, with many colonies at farms and villages on both Dartmoor and Exmoor, including one at 1,400 ft at Princetown. In the Postbridge area P. J. Dare considers it more numerous than the Swallow.

A recent enquiry covering only parts of Devon reported almost 2,000 nests in 1967 and concluded there had been no decrease during the three years 1965-7, although decreases have been reported in some previous years. The largest colony is at Exwell Barton, Powderham, which held about 150 occupied nests during these years and at least 134 in 1961. Other large colonies comprise one of ninety-three nests and three of 45-50 nests each.

Cliff nesting has occurred at a number of coastal sites including Watermouth (twenty-one nests in 1945), Ladram Bay (about twenty-one nests in 1946), and Wembury. These and others are listed in *BB* 33 : 18 and 38 : 134.

The first spring arrivals are frequently seen at the end of March but the bulk arrive during April. Prior to their departure in September and October, unusually large gatherings are occasionally reported, the most being about 5,000 at Exmouth on 24 September 1939 and the same number there on 9 September 1951. Most years a few late birds are observed in mid-November and occasionally December.

Spring and autumn passage migrants pass regularly through Lundy with the main numbers in April/May and September/October. A peak of 250 occurred on 22 May 1959, but usually less than 100 are counted in any one day.

SAND MARTIN *Riparia riparia*
Summer visitor, breeds

As a breeding species the Sand Martin is local though not uncommon, and nests in colonies mainly in sand pits and river banks, chiefly in east Devon. There has been no noticeable change of status.

In the reports of an enquiry into its status and distribution in 1962-5, P. W. Ellicott recorded concentrations on or near the rivers Axe, Otter, Exe, and Culm, and several small colonies on the Teign. In the north, small colonies exist at several points on the Taw, Yeo, Little Dart, and Torridge, and it has bred at Wistlandpound reservoir. It is scarce in the extreme west but bred near Lifton in 1964 at least. Sporadic breeding has been recorded at several localities on Dartmoor, including Powder Mills 1965-7, near Two Bridges, the Plym above Cadover Bridge, Manaton, and elsewhere, while at Lee Moor it breeds regularly. On the south coast a few pairs nest in sea cliffs at Thurlestone, Challaborough, and Dawlish.

The breeding population was estimated as just over 1,000 pairs in 1962, including three colonies of over 100 pairs. Marked fluctuations occur and the sites are often changed after a few years.

An early migrant, the first birds arrive in late March, with passage continuing into May and return migration occurring between about mid-July to early October. Records of more than usual include 1,000 at Slapton Ley in September 1947, 1,500 on the Exe on 6 May 1956 and 1,000 on 13 April 1966, and 3,000 at Slapton Ley in August 1968.

Sand Martins pass regularly through Lundy in spring and autumn. Peak numbers recorded include 400 on 4 May 1959 and 500 on 18 September 1961, but are normally much less. One ringed on Lundy on 30 August 1961 was retrapped three days later at Slapton.

GOLDEN ORIOLE *Oriolus oriolus*
Irregular spring passage migrant, has bred

D & M admit about a dozen occurrences of this species during the nineteenth century and, in addition, state that a pair bred at Piltor Abbey, Barnstaple, about 1865. Loyd, in *BSED*, quoted the *Field* of 10 February 1866 as stating that a pair nested near Totnes in 1865; if so, it is strange that this escaped the notice of D'Urban. According to the *Devon* and *Lundy Reports*, about fifty Golden Orioles have occurred during twenty-five of the years from 1928 to 1967, of which about twenty-four were reported from Lundy since 1947. Most of these records refer to singles, but pairs have occasionally been seen, eg at Dartmouth on 25 May 1931, and at Newton Abbot from 12 June to 22 July 1950. At least four were seen on Lundy during April and May 1964, and during the period from May to July 1965, and five were seen in May 1967. Except for two or three records each in July and September, all the occurrences since 1928 have been in April (10), May (22), and June (11). Of the mainland records, most refer to localities on or near the south coast, but at least three are from the fringes of Dartmoor, including Shaugh Prior, North Bovey, and Lustleigh, while three or so are from the north of the county, Hartland and Torrington.

Although the exact locality is not stated, and the record is not mentioned in the *Devon Reports*, it is recorded on p 181 of the *Report on Dorset Birds 1951* that a pair nested successfully in 1939 just on the Devon side of the boundary with Dorset. The nest is described and the two young were seen between 10 and 30 June 1939. The editors of this *Report* added that the details had only recently been received and that the record was absolutely reliable. In *BB* 60:273 the year is given as 1951 instead of 1939.

RAVEN   *Corvus corax*

Resident, breeds

The Raven, according to D & M, decreased very considerably during the second half of the nineteenth century and had become scarce in most parts of the county, but by 1906 D'Urban reported an increase which has since continued in practically all areas.

At the present time it is resident and widely distributed, breeding on the cliffs of both coasts, and, inland, in trees, quarries, and occasionally buildings. On Dartmoor it is thinly but widely distributed, with nesting records from many scattered localities, including some which, as in other parts of the county, have been in continuous use for twenty or thirty years. The *Devon Report* for 1967 recorded that during the 1960s it had nested in two thirds of all the 10 km squares of the county.

Page 208:
The west
coast of
Lundy.
Four species
of gulls,
three of auks,
and Fulmars
and Ravens
breed on the
precipitous
cliffs

In *BB* 49:28-31 H. G. Hurrell describes the establishment of a roost in his wood at Wrangaton, where up to eighty-six non-breeding Ravens roosted between January and April 1955. Large gatherings, which continue in varying numbers throughout the year, have been regularly reported at Winkleigh since 1960, when ninety-five were recorded in November, followed by 111 in November 1961, 130 in January 1962, and over 100 on several subsequent occasions.

A few resident pairs have bred regularly on Lundy since at least the 1880s, the breeding population being recorded as two pairs in 1914, four in 1930, and three in 1939. From two to four pairs in 1947-65, the number increased to at least nine breeding pairs in 1966. P. Davis states that most of the young leave the island in their first autumn. A flock of twenty-four was recorded on 25 March 1966.

CARRION CROW   *Corvus corone*
Resident, breeds

An exceedingly numerous and widespread bird, which has increased very considerably during the present century, the Carrion Crow occurs and breeds everywhere throughout the county from city parks to the heart of Dartmoor. In the West Webburn valley on Dartmoor, where the density is particularly high, P. J. Dare considers there are up to twenty breeding pairs per square mile.

There are many records of gatherings of well over 100 birds: G. H. Gush recorded 224 feeding on the Teign estuary in April 1952; H. S. Joyce noted a roosting flock of about 200 near Barnstaple in October 1953; and about 150 were seen at Winkleigh in November 1961. In June 1966, 195 were counted on the Teign estuary, and 112 at Topsham in July 1967.

In addition to the five to ten pairs which breed annually on Lundy, varying numbers of up to fifty or sixty visit the island, chiefly during spring and autumn, but also during the summer months.

HOODED CROW   *Corvus cornix*
Irregular visitor

D & M remark of this species that 'it appears to have been formerly much more numerous, frequenting the sea-shore near Plymouth in

N

winter, and Dr E. Moore says it was common about our coasts in winter. . . . We can remember when the Hooded Crow was a common autumn visitor to N. Devon, where we have seen numbers of them at the mouth of the Taw . . . (they) gradually became scarce and are now very rarely seen'.

The decline seems to have continued up to the present time, for the Hooded Crow nowadays occurs only as an occasional visitor, chiefly during the winter, but also as a spring migrant, and particularly so on Lundy where all the recent occurrences have been in the months of March to June, with none at all during the winter. The mainland records, with few exceptions, relate to coastal districts, there being none for Dartmoor and only one or two for other inland localities. The approximately forty recorded occurrences of this species during the present century up to 1967 include fourteen or fifteen for Lundy; four for Branscombe, Seaton, and Exmouth, mentioned by Loyd; seventeen listed in the *Devon Reports*; and a further three contained in D'Urban's MS. These last relate to a single bird seen at Newport House, Countess Wear on 4 January 1900, one shot on the Taw estuary near Barnstaple on 7 November 1906, and one seen at the mouth of the River Otter on 5 March 1911.

Of the seventeen records in the *Devon Reports* from 1928 to 1967, those that merit particular mention refer to a bird that was present at Nymet Rowland during the winter of 1933-4, being last seen on 6 March 1934; one seen at Tawstock on 12 April 1942; one which remained about Southpool Creek near Salcombe during the autumn and winter of 1946 and the whole of 1947, and was joined by a second bird on 30 December 1947; one on Northam Burrows from 5 to 21 November 1947; one inland at Meeth on 29 December 1954; one observed on the Exe estuary from 4 February until 11 March 1956; one at Axmouth on 7 June 1962; and two at Brixham on 2 July 1964.

ROOK   *Corvus frugilegus*

Resident, breeds

D & M described the Rook as generally distributed, very abundant, and, in some places, increasing enormously. A resident species with no evidence of migratory movement, it is extremely widespread as a breeding bird, with rookeries in practically every 10 km square, except for much of Dartmoor and the high open parts of Devon Exmoor. On the latter there is a small colony of sixteen nests at Blackmore Gate, and on Dartmoor a number of small colonies of up to about twenty nests have been reported, including those at Two Bridges, Postbridge, Fernworthy, and Broadford. Probably the highest is the one at 1,450 ft at Princetown.

Foraging flocks occur in the remotest parts of the moor during summer, when they have been observed in the Cranmere Pool area, but for the most part they desert the moor from about November to February.

No complete census of rookeries appears to have been made, but a partial survey conducted by Dr I. E. P. Taylor in 1966, covering possibly about a quarter or third of the county, gave a total of 224 colonies containing 4,583 nests. The largest number of rookeries in any one 10 km square was fifty-nine, in the Silverton area, and the largest rookeries were of 125 nests at Killerton and 132 at Tiverton. Three-quarters of the sites were at or below 400 ft.

One of the largest winter roosts reported was that of about 6,000 birds at Netherton during January 1950.

Ones and twos are recorded annually on Lundy, mainly from March to May. They have been observed in all months except December and February, with maximum numbers of twenty in September 1930 and twelve on 8 April 1953.

JACKDAW *Corvus monedula*
Resident, breeds

An abundant resident, the Jackdaw breeds in great numbers on the coastal cliffs, particularly in the north, but is also widespread throughout the county, nesting in old trees, church towers, farm buildings, quarries, and old mine-workings. As a breeding bird it is restricted on Dartmoor by the availability of nesting sites, but breeds at a number of localities such as Vitifer mines, Powder Mills, and other derelict buildings.

Foraging flocks of up to 1,000 birds occur regularly on grassland, and 2,000 were recorded at Colaton Raleigh in August 1952. Large roosts include one of about 3,000 at Netherton in March 1953, and 1,000-2,000 in conifers at Burrator in February 1962.

Singles and small parties of up to about six occur fairly regularly on Lundy, mainly in February to May and September to November, but they have been recorded in all months. Up to fifty-three were present during April and May 1948, twenty-seven in November 1952, and 'large parties' wintered in several years during the 1930s. Despite the apparent suitability of the cliffs as nesting sites, the Jackdaw does not breed on the island, although it was said to around 1908.

MAGPIE   *Pica pica*
Resident, breeds

The Magpie, which is now abundant, extremely widespread, and breeds in all but treeless and hedgeless areas, has increased during the present century. In 1906 D'Urban feared that it was gradually decreasing in some districts, but the trend since 1928, if not earlier, has been one of increase.

On Dartmoor it is absent only from the treeless areas, and in 1945 G. M. Spooner stated that it occurred well above the highest levels of cultivation, while in 1956 P. J. Dare considered it to be very numerous in the Postbridge area. Two were noted at Black Tor Copse in May 1965. The *Devon Report* for 1963 noted only a slight reduction in numbers as a result of the particularly severe winter of 1962-3. In the Axe area R. Cottrill recorded over fifty pairs in 50 square km in 1966.

For Lundy, Davis gives eight records since 1887, the two most recent being singles 12-18 June 1952 and 2 July 1953.

NUTCRACKER   *Nucifraga caryocatactes*
Vagrant

Although the Nutcracker has been reported about eleven times in Devon, only a few of the occurrences can be regarded as completely satisfactory. Of the seven instances enumerated by D & M, most lack either the date or locality, and some have neither. Perhaps the two most reliable of the old records are those of a bird shot in north Devon in August 1808, and another killed at Washford Pyne Moor near Tiverton in 1829.

There are four records for the present century, but even these are not all fully substantiated. The first relates to a bird observed at Budleigh Salterton on 27 December 1938. In November 1944 one frequented some gardens at Preston, Paignton, for several days. It was eventually found dead and is now preserved at the Torquay Museum. As in the case of the majority of Nutcrackers visiting Britain, this specimen is of the Slender-billed race and is the only Devonshire example that has been sub-specifically determined.

The latest occurrence refers to a bird, thought to be of the Slender-billed form, which was watched by S. D. Gibbard and W. N. Bolderston at Trusham on 6 September 1945 and is described in the *Devon Report* for that year. The Nutcracker has not been recorded on Lundy.

Since writing this account, five were recorded in different parts of the county during the autumn of 1968.

JAY   *Garrulus glandarius*

Resident, breeds

D'Urban, in the *VCH*, noted that the Jay had become scarce in many areas through persecution, but was still fairly common in some localities. With the gradual cessation of persecution during the past 30-40 years, it has increased and is now widespread throughout the county in all wooded areas, except on high ground. On Dartmoor, although it breeds in small numbers in the conifer plantations at Bellever and is common in the woodlands fringing the moor, it is scarce in the Postbridge district. At lower altitudes it has benefitted by afforestation and is plentiful in all the plantations.

An influx in the autumn has been noted in several recent years, particularly in October 1957 when movements of flocks of up to eighteen birds were observed in several different parts of the county. This irregular movement is borne out by the occurrence of a single bird on Lundy on 26 September 1965, constituting the first and only record for the island.

CHOUGH   *Pyrrhocorax pyrrhocorax*

Formerly resident, now vagrant

D'Urban, in the *VCH*, writes: 'Resident in small numbers in some spots on the north coast of the county. It is now unfortunately extinct on Lundy Island where it was formerly numerous'. The

Chough has long since ceased to breed in Devon, where it is last known to have nested in 1910, according to D'Urban's MS. It now occurs here only as a rare vagrant.

In D & M's time it was a familiar bird along practically the entire north coast, including Lundy, and occurred at places on the south coast; but by the turn of the century it had already deserted many of the localities which had formerly known it. According to Loyd, it bred on Lundy until about 1890, since when it has been reliably recorded there on only two occasions, a single bird having been present from 18 to 24 October 1949, and one from 20 February until 3 March 1952.

The Chough seems to have survived on the north Devon coast until about 1910, in which year it nested on the precipitous cliffs at Lynton, according to D'Urban's informant, a James Turner, who knew of one, and possibly two, nests at Lynton and another further east on the Somerset coast. One of the last breeding records for the south coast was at Berry Head where the eggs were taken in 1880.

Of nine occurrences listed in the *Devon Reports* from 1930 onwards, eight refer to single birds, and one to a party of three seen at Berry Pomeroy on 6 April 1946 by a person said to be familiar with the species. The records of single birds refer to Bolt Head in April 1930; Hartland Point, one seen many times during the autumn of 1930; Ilfracombe, one said to have been seen in 1931; Hartland Point, one reported caught in a rabbit trap in 1939; Budleigh Salterton, 14 August 1943; between Shaldon and Maidencombe, 28 August 1943; Morte Point, one closely observed between 6 and 20 November 1949 by Miss V. M. Bury; and lastly, a bird that may possibly have escaped from captivity, seen near Crediton by F. R. Smith on 20 February 1957.

The last pair of Choughs in Cornwall lingered on until 1967, when one was killed; the other still survives at the time of writing.

GREAT TIT   *Parus major*

Resident, breeds

An abundant resident species, the Great Tit is widely distributed in woodlands, copses, hedgerows, and gardens throughout the county. It occurs in the cultivated and afforested parts of Dartmoor but is absent from open moorland. The numbers do not appear to have changed appreciably since the time of D & M, and it withstood the very severe winter of 1963 better than many species. Although mainly sedentary, some autumn passage has occasionally been noted in coastal localities.

Although the Great Tit does not breed on Lundy, a small autumn movement has been detected in almost half of the past 20 years, and a smaller spring movement in some years, but the species is not recorded annually and has only very exceptionally been known to winter on the island.

BLUE TIT   *Parus caeruleus*

Resident, breeds

By far the commonest of the tits, this species is very widespread and is an abundant breeding bird throughout Devon, except on the more elevated and exposed parts of Dartmoor and Exmoor. That it sometimes occurs on high open ground, however, is evidenced by a record of four, possibly migrants, at 1,240 ft, on the top of Ugborough Beacon on 11 November 1951.

Although there is some fluctuation in numbers, with a decrease in hard winters, the species quickly recovers and its status has apparently not changed since last century.

The Blue Tit does not breed on Lundy, and occurs there only occasionally, having been recorded in autumn and spring in only about 8 of the past 20 years. The numbers do not normally exceed about six, but they were exceptional in the autumn of 1957 when a peak of about eighty was recorded on the island on 4 October, followed by fluctuating but smaller numbers during the rest of the month. A single bird was recorded as wintering on Lundy in 1962.

A Blue Tit ringed at Haslemere, Surrey in December 1962 was recovered at Honiton in November 1963; there are similar recoveries from Somerset and Dorset.

COAL TIT   *Parus ater*

Resident, breeds

Described by D & M as tolerably common, it is probable that the Coal Tit has since increased, due to the afforestation with conifers of large areas of country previously unsuited to its requirements. Although not so numerous as either the Great or Blue Tits, it is nevertheless a common and widespread species and is particularly numerous in the conifer plantations throughout the county. Resident on Dartmoor, it is restricted to woodland and occurs principally in the plantations.

That there is some migratory movement is shown by the occasional occurrence of this species on Lundy, where it has been recorded in about 8 of the past 20 years, usually during October and November, and very occasionally in the spring. In the exceptional movement of tits during the autumn of 1957, thirty were recorded on the island on 11 October and smaller numbers on the few previous days. Up to six wintered in 1957-8 and were last seen on 30 April.

CRESTED TIT   *Parus cristatus*

Vagrant

Although the Crested Tit has been recorded six times in Devon, few of the occurrences can be considered reliable, as in most cases the data are incomplete and the records lack conviction. D & M knew of only one occurrence, a sight-record of a bird observed at Torquay on 26 March 1874 and reported in the *Zoologist* for that year. According to H. Coates's *Catalogue of Devonshire Birds in the Torquay Museum*, an example which is said to have been taken at Kingskerswell some time prior to 1891 was presented to the Museum in 1931.

The *Devon Reports* contain three records, the first of which is a second-hand report of one seen on two occasions in a garden at Bridford during March 1938, but this, like the next record, is inadequately authenticated. The second record, also reported at second hand, refers to one supposed to have been seen at Paignton on 24 May 1945, but again lacks details of identification. The only convincing record of these three is that of a bird which came to the window-sill of Mrs Fordred Neale's house at Torquay, and was observed twice on 25 and once on 27 January 1945. This bird took food from the window-sill and was thus observed at very close quarters.

Lastly there is the record in *BB* 43 : 118 of one closely observed on the cliff edge between Dawlish and the Warren by T. R. F. and L. V. A. Nonweiler on 28 December 1947, and described in such detail as to leave no doubt of its correct identification; this record, however, is not in the *Devon Report*.

MARSH TIT   *Parus palustris*
Resident, breeds

Widely distributed throughout the county, the Marsh Tit is a common resident in deciduous woods, small copses, rough hedgerows, and suitable gardens, and feeds regularly at bird tables. It is frequent in the wooded river valleys from the fringe of Dartmoor downwards but is absent from open, exposed moorland, although it occurs occasionally and a very few pairs breed in parts of the semi-cultivated areas around Postbridge.

It was not so badly affected as many species by the severe winter of 1962-3, but the numbers show some fluctuation between hard and mild winters. There is no evidence of migration, but a single bird was recorded on Lundy on 15 and 18 January 1958, constituting the only record for the island. There appears to have been no change of status since the last century.

WILLOW TIT   *Parus montanus*
Resident, breeds

The Willow Tit, being unrecognised in Britain until the close of the nineteenth century, was unknown to D & M and is consequently not mentioned in either *The Birds of Devon* or even in the subsequent *VCH*. The first field record for Devon appears to be that of a pair reported on Woodbury Common in May 1928 by W. Walmesley White (*BB* 22 : 38), although the species has not since been regularly observed, if at all, in this district.

Writing in 1929, Loyd (*BSED*) stated that nests had been found near Seaton and at Branscombe, but gave neither dates nor any supporting evidence, although more recent records indicate that the species does occur sparingly in east Devon. Satisfactory proof of breeding in Devon was not forthcoming until 1943 when B. G. Lampard-Vachell located two pairs at Weare Giffard and closely watched one of them excavating its nesting hole in an alder stump.

In the following year he had two pairs under observation in the same locality, where breeding was again reported in 1946 and 1947.

Whilst single birds and the odd pair have been recorded from many parts of the county, and it does occur in several of the river valleys, its stronghold lies in north-west Devon between the heavily wooded Torridge valley and Tamar Lake. In this area it is nowadays recorded annually from many wooded localities in which it co-exists with the much more plentiful Marsh Tit, and, although it is far from common, the Willow Tit can usually be found in suitable habitats.

That it wanders to some extent in winter is suggested by two records from Dartmoor, from which it is normally absent. The first is of a single bird seen by P. J. Dare on 12 January 1958 in the West Webburn valley, and the second, also a single bird, was observed by H. G. Hurrell at Wrangaton on 1 November 1964, being his first record there in 25 years. From the south it appears to be entirely absent except for the occasional bird in the Teign valley. In north and mid-Devon four were seen at Wistlandpound and five at Lapford during January 1967.

It has not been recorded on Lundy.

LONG-TAILED TIT   *Aegithalos caudatus*

Resident, breeds

Normally an abundant and widespread species, its numbers were drastically reduced by the hard winters of 1939-40, 1946-7 and 1962-3 (and earlier ones), but in each case it recovered within a few years. It is generally distributed throughout the county except on high moorland, but occurs and a few pairs breed in areas with suitable cover at about 1,000 ft around Postbridge, near which a flock of forty-five was recorded in June 1957.

The *Devon Report* for 1960 stated that partly due to an influx it was very numerous everywhere in the autumn and winter, and many flocks were recorded in coastal areas, while in 1967 it reported that records from over 100 localities indicated that the numbers were higher than in 1962.

Further evidence of migratory movement is provided by the occasional occurrence in autumn and more rarely spring of this species on Lundy, where it has been reported in 7 years since 1932,

with a maximum of twelve birds in October 1958.
There is no evidence of any long-term change of status.

BEARDED TIT   *Panurus biarmicus*
Vagrant

The history of the Bearded Tit in Devon is rather obscure and there
are no properly authenticated breeding records, although several
places are mentioned where it is said to have occurred during the
nineteenth century. D & M wrote that it was 'a casual visitor of
very rare occurrence at the present day, but appears to have been a
resident about fifty or sixty years ago'. Mathew himself records
having seen a flock of about a dozen one autumn day on a marsh
near Barnstaple, but it now seems most likely that they had 'erupted'
from elsewhere.

According to Dr E. Moore in 1830 this species was formerly resident in the reed beds at Topsham, and is said to have been met with
on the Exe at Thorverton and also near Bovey Tracey. Howard
Saunders evidently referred to Slapton Ley when he wrote that it
bred at one locality in Devon, but D & M were able to add nothing
further to this except that a Mr Bower Scott of Chudleigh assured
them that he actually saw a Bearded Tit at Slapton around 1872.
Although there are no definite records of its breeding at Slapton
during the present century, or even occurring there until recently,
it was still presumed to do so by T. A. Coward writing in 1920, and
by Kirkman and Jourdain writing in 1930. The matter was ultimately
settled to H. F. Witherby's satisfaction in 1932 by an intensive but
unrewarding search, about which he later wrote to me as follows:
'I went into this with some care a few years ago and watched for it
myself, as did also E. M. Nicholson and W. Walmesley White who
both spent considerable time over it. I do not think the bird still
breeds there although it did in the old days'.

Positive proof of its occurrence as a migrant at Slapton was obtained in the autumn of 1965 when, following an impressive eruption
of Bearded Tits from Holland, they were recorded over much of
Britain, including a number of localities in Devon (H. E. Axell in
BB 59:513-43). The first four birds were seen on the Otter Marshes
on 13 October, followed by twenty-five on 14 and 15. Up to twelve
were seen on Dawlish Warren during late October and early Novem-

ber, eight at Slapton Ley from 16 to 28 October, of which two were trapped and ringed. Smaller numbers were reported from the estuaries of the Axe, Tavy, and Taw. Unfortunately the birds gradually trickled away, but fifteen wintered on the Otter Marshes, where the last four left on 4 March 1966; fifteen remained on the Exeter Canal until well into January; and two were reported at Tamar Lake on 13 February and the last one was seen there on 9 April. A pair of these delightful birds occurred at Dawlish Warren in the following year, being observed by F. R. Smith on 25 November 1967.

NUTHATCH   *Sitta europaea*

Resident, breeds

The resident and sedentary Nuthatch, although by no means abundant, is nevertheless tolerably common and occurs throughout the county wherever there are old deciduous trees, whether in suburban gardens, parkland, or isolated copses on moorland. Although absent from the open moorland, it occurs at 1,200 ft on Dartmoor, a pair having been located at Archerton and other localities in the Postbridge district. Similarly it occurs in suitable habitats on west Exmoor. It has been recorded as breeding in all parts of the county, and an indication of its numbers is given by R. Cottrill, who in 1966 recorded about twenty-three pairs in an area of 50 square km bordering the River Axe. Following the severe winter of 1962-3, it was described in the *Devon Report* for 1964 as being plentiful in all areas.

The Nuthatch has not been recorded on Lundy; and the only evidence of any movement is the report of a single bird seen to fly in from the sea near Start Point on 29 August 1964. If there has been any change of status since D & M's time, it is probably a slight increase.

TREE CREEPER   *Certhia familiaris*

Resident, breeds

A common resident species, the Tree Creeper occurs and breeds throughout the county, in woodlands and parkland, and in gardens and town parks with old trees. It occurs as a scarce resident in suitable habitats on Dartmoor, where it was recorded as breeding in 1962, and as occurring at 1,100 ft at Cator in April 1967, in addition to other records for the Postbridge area during the 1950s.

The *Devon Reports* contain breeding records for localities in all parts of Devon, and except for slight fluctuations due to hard winters the numbers appear to be unchanged since D & M's time. In 1966 R. Cottrill recorded thirty-two pairs in an area of 50 square km bordering the River Axe. A bird thought to be of the Northern race (*C.f.familiaris*) was observed by F. R. Smith and A. V. Smith at Powderham on 27 February 1955.

The Tree Creeper has been recorded about twelve times on Lundy in eight of the years since 1950. All the records refer to single birds which occurred during the months from July to November.

WREN *Troglodytes troglodytes*
Resident, breeds

It is doubtful whether any other species has such a wide distribution in Devon as the Wren, which occurs wherever there is a small amount of low cover. An abundant resident, it is found from the coastal cliffs and shores to the highest parts of Dartmoor, and is equally at home on the boulder-strewn hillsides or the lowland reed beds as about the farmyards, lanes, and suburban gardens.

This truly ubiquitous species breeds commonly on Lundy, where it is resident. The breeding population was estimated by Perry to be thirty-five pairs in 1939, and by Allen as fifty to sixty pairs in 1944. The *Lundy Reports* give its status as widespread and abundant, with probably a small autumn passage during October and November. Two birds seen on 14 November 1954 were tentatively assigned to the race *T.t.kabylorum* which inhabits Spain and north-west Africa.

The population of Wrens was badly reduced by the severe winter of 1962-3, both on Lundy and the mainland, and particularly on Dartmoor, but the numbers had completely recovered by the end of 1966 and increased even further during 1967. There is no evidence of any change of status since the last century.

DIPPER *Cinclus cinclus*
Resident, breeds

The Dipper is well distributed in Devon, occurring on almost every fast-running river and stream, and ranging from about 1,200 ft on

Dartmoor down to sea level, while on some of the smaller rivers such as the Lyn and Otter it breeds within a short distance of the sea. Although it is normally absent from the estuaries, there are reports of winter occurrences on most of them, including one on the Exe at Starcross in December 1958.

Some idea of the numbers may be gained from the *Devon Report* for 1950 in which it is recorded that a pair occupied every ¾ mile of the Torridge from Bulkworthy to Merton, a distance of approximately 20 miles, and from Dare and Hamilton's figure of five to ten breeding pairs in an area of 15 square miles around Postbridge.

The Dipper seemed well able to withstand the extreme cold of 1963, as there were no reports of any significant reduction in numbers. It appears, indeed, that it has increased since the close of the last century, when D'Urban reported that, although formerly common, it had become greatly reduced in numbers.

Although a single bird was recently recorded on the coast at Slapton in the autumn, there is no record of any regular migratory movement and it has not been reliably reported from Lundy.

MISTLE THRUSH   *Turdus viscivorus*

Resident, breeds

The Mistle Thrush is common and widely distributed, breeding in city parks, large gardens, orchards, woodlands, and conifer plantations throughout the county, including suitable areas on Exmoor and Dartmoor. Although subject to losses due to severe winters such as 1962-3, the numbers are restored within a few years.

From July onwards many breeding localities are deserted and flocks concentrate in open country, and are frequent on open moorland on both Exmoor and Dartmoor where flocks of fifty or more are regularly reported; a large flock of 100 was recorded at Postbridge on 9 August 1956. In winter the numbers are probably augmented by birds from more northerly counties, as some movement is noted, eg a flock of about 150 flying west over Woodbury Common on 2 October 1946 and a great passage reported by Eagle Clarke at the Eddystone Light on 12 October 1901. The evidence from Lundy, however, is of only very light spring and autumn passage, the most recorded being a flock of twenty on 15 November 1958.

A pair bred on the island in most years from 1929 to 1941, but breeding has not since been recorded.

FIELDFARE *Turdus pilaris*
Winter visitor

A regular winter visitor in fluctuating numbers, the Fieldfare is abundant in some years but comparatively scarce in others. It occurs in flocks of usually under 100 but not infrequently 300-400, mainly in open country, on pastures and rough grazing land, whether at sea level or on high ground on Dartmoor and Exmoor, and is well distributed throughout Devon.

The first arrivals are reported in October, occasionally late September, and flocks are present until the end of March, frequently into mid-April and occasionally early May. Weather movements often occur during December and January, when flocks totalling thousands of birds are recorded passing westwards along the coast. One such movement occurred on 9 December 1967, when 1,000 passed westwards over Exeter and an estimated 8,000 were reported flying northwards at Staddon Heights, presumably following the coastline.

Large roosts are occasionally recorded, eg 2,000 at Killerton during February 1953, such roosts sometimes being shared with Redwings. Another unusually large flock comprised 2,000 birds at Soussons in February 1967, and a migratory movement over the centre of Dartmoor totalled some 2,000 birds passing north-east over Lynch Tor on 10 April 1966.

Variable numbers occur regularly at Lundy on spring and autumn passage, when peaks of up to about 300 are occasionally recorded during November. Small numbers are frequently present in all months from October to March.

SONG THRUSH *Turdus Philomelos*
Resident and winter visitor, breeds

The Song Thrush, described by D & M as abundant, is now far less plentiful than the Blackbird, having been very seriously reduced by the hard winter of 1963, and possibly 1947. It is still recovering from the effects of 1963 and is again widespread and tolerably common, except on moorland, but in nothing like its former numbers. As a guide to its relative abundance, over four times as many Blackbirds as Song Thrushes have been ringed on Lundy between 1947 and 1966, the figures being 1,115 and 244. On Dartmoor it is resident in culti-vated areas and conifer plantations but is not numerous.

Examples of weather movements in winter are a south-westerly movement observed at various points along the south coast in January 1960, and a movement of about 600 noted at Stoke Point on 10 December 1967. It is generally assumed that there is some immi-gration of Continental birds in winter, and one or two trapped on Lundy were thought to be of this race.

Varying numbers are reported on spring and autumn passage on Lundy, with a peak of 200 on 27 October 1960, and an exceptionally heavy spring passage of 700 on 11 March 1962, but in some years only a very few are recorded. Davis records that it bred regularly until 1943, with a maximum of nine pairs in 1930. One or two pairs again bred from 1948-51 and 1957-62, with a maximum of three or four pairs in 1962, but none since.

REDWING   *Turdus iliacus*

Winter visitor

The status of the Redwing, a generally abundant winter visitor, but in fluctuating numbers, does not appear to have changed since D & M's time. Although usually less plentiful in mild winters, it is normally well distributed on grassland throughout the county and occurs commonly on Dartmoor, but moves to lower ground in severe weather. Arriving in about mid-October, occasionally in late Sep-tember, it is present until late March or early April. With the onset of hard weather, extensive movements are noted, involving flocks sometimes of several thousand birds. Prolonged cold drives them into town parks and gardens, and in severe winters, such as 1962-3, many perish.

Large roosts are formed in different areas, eg 2,000 in rhododen-drons at Killerton in January 1950 and several subsequent years, several thousand at Chawleigh in 1950, 6,000 at Coombe Brake,

Woodbury in February and 6,500 at Peamore during February and March 1955, and 4,000 on the Erme in December 1967.

On Lundy it occurs as a winter visitor and abundantly as a spring and autumn passage migrant, with peaks of up to 500, often much less, in October and November, and an unusually large number of 2,000 on 11 March 1962. A bird of the Icelandic race (*T.i. coburni*) was trapped on 6 November 1951, and one seen on 12 and 13 May 1964 was considered to be of this race. A bird ringed on Lundy in November 1958 was subsequently recovered in Karelia, USSR, in August 1960.

Several Redwings ringed as young in Finland and Sweden have been recovered in Devon in winter.

RING OUZEL   *Turdus torquatus*
Summer visitor, breeds

Although the Ring Ouzel is evidently not now as plentiful on Dartmoor as it appears to have been in D & M's time, and has certainly decreased on western Exmoor, it is still well distributed over the northern half of Dartmoor, where breeding pairs have been reported in many different localities during the past 10 years. Nesting pairs are mostly located at altitudes over 1,250 ft and occur in many remote parts of the moor, almost to Cranmere Pool. In 1959 E. H. Ware recorded at least fifteen breeding pairs in one particular locality, where the species still breeds annually but in slightly smaller numbers.

In north Devon records of nesting on western Exmoor are nowadays very few, but breeding was reported on Brendon Common in 1960 and several birds were noted at a locality on western Exmoor in May 1959, while during the 1930s and 1940s there were reports of nesting in the Badgworthy Water area. It seems probable that a pair or two still breed on Devon Exmoor.

An early migrant, the first arrivals are noted at the end of March and beginning of April and departing birds are observed on the coast during October and occasionally early November. Some recent records of autumn flocks on Dartmoor include twenty-five at Powder Mills in October 1956, about twenty near Combestone Tor in September 1964, sixteen at Burrator in October 1966, and twenty-eight at Combestone Tor on 22 October 1967.

o

A regular passage migrant on Lundy, it is recorded in both spring and autumn, with peaks of up to about twenty and a maximum of twenty-three on 7 April 1966.

BLACKBIRD   *Turdus merula*

Resident and winter visitor, breeds

Abundant and widely distributed throughout the county, the Blackbird is resident and breeds commonly in all suitable localities. It outnumbers the Song Thrush and is better able to withstand severe winters, as was particularly noticeable in the winter of 1962-3, when the Blackbird suffered much less than many other species. As a breeding bird, it is absent only from the open moorland in the highest central part of Dartmoor, but occurs commonly in the fir plantations and in the indigenous upland woodlands such as Wistman's Wood. In the Postbridge area, which may be regarded as typical of much of Dartmoor, P. J. Dare found it resident and breeding in all suitable habitats, and mentions a pair nesting at over 1,250 ft at Headland Warren Farm, where it overlaps with the Ring Ouzel and shares the same feeding ground.

On Lundy the Blackbird is present throughout the year, the twelve to fifteen breeding pairs being augmented by both spring and autumn passage migrants and irregular cold weather movements. A heavier than usual autumn passage was recorded in 1958, when a maximum of about 500 birds were present on 24 October.

D & M described this species as resident, generally distributed, and abundant, and remarked that it was increasing, a trend that is apparently still continuing at the present time.

AMERICAN ROBIN   *Turdus migratorius*

Vagrant

An American Robin, an addition to the British List, occurred on Lundy during a period of strong westerly winds in October 1952 and was present on the island from 27 (possibly 25) October until 8 November. It was trapped, ringed, photographed, and critically examined by Peter Davis on 27 October and was considered to be a first-winter bird, probably of the Eastern race (*Turdus m. migratorius*).

During most of its stay it remained on the open grassland on the top of the island where it associated with Redwings and Blackbirds, and was observed feeding principally on earthworms. A full account of the occurrence, together with an assessment of the meteorological conditions, which brought several other American species to the British Isles and accounted for the catastrophic 'wreck' of Leach's Petrels, is given by Peter Davis in *BB* 46 : 364-8.

Another example occurred at Braunton Burrows on 29 October 1955, where it was studied by A. S. Cutcliffe, whose detailed description is included in the *Devon Report* for 1955. The third occurrence relates to a bird seen on Lundy on 7 November 1962, for which the record was accepted by the Rarities Committee in *BB* 56 : 403.

WHITE'S THRUSH   *Zoothera dauma*
Vagrant

A rare vagrant from Asia, White's Thrush has been reliably reported on only two occasions, in addition to which it is listed by Chanter as a doubtful visitor to Lundy, where it has since been definitely recorded. One of these large thrushes in company with three or four other birds, apparently of the same species, was shot in Dean Wood near Ashburton on 11 January 1881 and, as D & M relate, it was at first mistaken for a Woodcock when flushed.

The other fully authenticated occurrence refers to a single bird which was present on Lundy from 15 October until 8 November 1952, during which time it remained in the small dense wood in Millcombe. In his account in *BB* 46 : 455 Peter Davis states that the bird was extremely shy and was difficult to watch because it flew at once into cover on being disturbed.

ROCK THRUSH   *Monticola saxatilis*
Vagrant

An addition to D & M, the Rock Thrush is a summer visitor to central and southern Europe and Asia, and up to the mid-1960s has been recorded in the British Isles on less than ten occasions. The only record for Devon refers to an adult male which was seen at the Eddystone Lighthouse by H. S. Taylor and others on 30 and 31 May 1963. This record did not come to the notice of the editor of the *Devon Report* and is not mentioned in it, but was accepted by the Rarities Committee and published in *BB 57 : 272*.

WHEATEAR   *Oenanthe oenanthe*

Summer visitor, breeds

There appears to be no indication in Devon of the decrease in numbers of this species noted in several other southern counties. A summer visitor, arriving in late March and early April and remaining until about mid-October, sometimes later, the Wheatear occurs commonly on migration on both coasts and on the estuaries. As a breeding species it is common and widespread on Dartmoor where it occurs in all suitable rocky and stony areas, mainly above 1,100 ft. Dare and Hamilton state that 100-150 pairs breed in an area of 15 square miles around Postbridge.

D & M mentioned breeding at several coastal localities, which have long since been deserted, but sporadic breeding has been recorded at a number of sites since 1930, including Start Point, Dowlands Landslip, Woodbury Common, Haldon, probably at Hope's Nose, Wistlandpound, and Sherrycombe near Combe Martin. Several of these localities have been used during the 1960s, in addition to which it breeds regularly on Northam and Braunton Burrows, and most probably on parts of western Exmoor.

A small population of from three to ten pairs nests almost annually on Lundy, where the species occurs commonly on spring and autumn migration, with peaks of up to 100 during late April and early May and again in August and September. The Greenland race (*O.o.leucorrhoa*), which has been trapped and critically examined on a number of occasions, occurs in both spring and autumn. One which had been ringed as a juvenile in Greenland in 1958 was trapped on Lundy on 8 May 1959.

BLACK-EARED WHEATEAR   *Oenanthe hispanica*

Vagrant

A male Black-eared Wheatear of the black-throated form, the first record for the county, was identified by A. D. G. Smart at Musbury

Castle near Seaton on 3 May 1947, and was subsequently observed by A. W. L. Mayo and the Rev F. C. Butters, being last seen by the latter on 6 May. It was watched at close range and under good conditions, so it was possible to obtain a detailed field description, the full account of which is given in *BB* 40:345; but even so, it was not possible to assign it to a particular race.

On 11 May of the same year a male of the white-throated form was observed at Vitifer, Dartmoor; full details are given in the *Devon Report* for 1947.

STONECHAT   *Saxicola torquata*

Resident, breeds

A resident species, breeding on gorse-clad commons, moorland, and uncultivated land, particularly along the coast, the Stonechat suffers considerable mortality in severe winters. In recent years its numbers were badly reduced by the hard winters of 1946-7, 1955-6, and 1962-3, but the losses are usually made good within a few years. Further reductions, however, have been caused by the cultivation of marginal land and human disturbance, including heath fires.

A BTO survey conducted by J. D. Magee gave the breeding population of Devon in 1961 as ninety-six pairs, which may well have been far short of the actual number. The species normally breeds along much of the north and south coasts, on Exmoor and Dartmoor, Haldon, Woodbury, Chudleigh Heath, and elsewhere.

The *Devon Report* for 1967 stated that the numbers were almost back to the 1962 level, and mentioned fifteen pairs in June between Pickwell and Rockham Bay on the north coast, and family parties every few hundred yards along the 6-7 miles of coast from Welcombe Mouth to Hartland Point. In 1962 there were fifteen pairs between Hope Cove and Salcombe on the south coast. In an area of 15 square miles around Postbridge, P. J. Dare and L. I. Hamilton gave a maximum population of 10-15 pairs in the years 1956-67.

Most moorland breeding areas are deserted from about September to March, when the birds move to lower ground, mainly along the coast and estuaries. Some coastal movement is reported in most autumns and sometimes in spring, eg over forty at Stoke Point on 9 September 1967.

On Lundy, twenty-eight pairs were reported breeding in 1930, about fifteen in 1939, four or five in 1942, and one pair 1951-3. Four pairs were present in 1960, at least six pairs nested in 1962, but none since. Small numbers occur on spring and autumn passage and occasional birds winter on the island.

WHINCHAT   *Saxicola rubetra*
Summer visitor, breeds

A summer visitor, the Whinchat is mainly confined to Dartmoor and Exmoor, but breeds occasionally on other commons and moorlands, such as Woodbury Common, Haldon, and Chudleigh Heath, all of which have been used in recent years. Although D & M stated that it was not a numerous species, it is certainly not uncommon at the present time, being well distributed over much of Dartmoor, particularly in the combes, and many breeding pairs are recorded annually. In his account of the birds of the Postbridge district of Dartmoor, in *Devon Birds* 11:30, P. J. Dare describes this species as a 'common summer resident, widely distributed over all suitable ground and far outnumbering the Stonechat; particularly attracted to young, open conifer plantations; minimum population in the area east of the Dart in 1957 was fifty pairs'. In May 1966 E. Griffiths encountered twenty birds in a few square miles near Postbridge and Soussons. In north Devon A. J. Vickery has recorded it breeding at Wistland-pound in recent years, and when living in that part of the county I used regularly to find it breeding on the west side of Exmoor.

On migration the Whinchat is observed annually at many places along both coasts, particularly the south, but is much more plentiful during the autumn than the spring migration. Although the species visits Lundy on spring and autumn passage, the numbers are always small in the spring movement and rarely exceed four or five individuals, while in some years none at all are recorded at this season. On the autumn passage, however, it occurs regularly and in greater numbers, the movement lasting as a rule from late August until about the end of September, with the odd straggler in October or occasionally November. The number seen in any one day is not very large and the peak is rarely more than twenty. Although Loyd believed that June records for Lundy might refer to breeding birds, there is no record of the Whinchat ever having nested on the island.

REDSTART *Phoenicurus phoenicurus*

Summer visitor, breeds

A regular summer visitor, with an uneven breeding distribution, the Redstart has increased during the past 20 years or so and is slowly spreading westwards. A bird mainly of wooded and hilly country, it breeds in a number of scattered localities in east Devon : from the south of Exmoor along the Exe valley to Exeter, in several parishes around Crediton, and from Haldon westwards to Dartmoor, where it is widely distributed and, in some parts, quite common. In the north it breeds on Exmoor from the Somerset border westwards to Hunter's Inn, including much of the country around Brendon and Badgworthy Water. It also occurs in parts of the Taw valley, around Eggesford and Chulmleigh, while breeding was recorded at Tawstock in 1957. On the Torridge, where it was formerly absent, a singing male was recorded at Landcross in 1964, and on the north coast singing males have been reported in recent years at Clovelly. The most westerly breeding record is at St Giles on the Heath, where a pair bred in 1952; while breeding was suspected in the adjoining parish of Werrington in 1958.

The Redstart is absent from practically the whole of the north-west and also from the South Hams, but a pair bred at Newton Ferrers in 1967 and near Hartland in 1968. A four-year enquiry from 1957 to 1960, conducted by P. W. Ellicott and S. G. Madge, revealed a significant spread on the south and west of Dartmoor, since when a further increase in numbers has been noted in many parts of Dartmoor. Breeding was recorded at Wrangaton for the first time in 1961, and the species has since spread to other woods on the southern fringe of Dartmoor, while in the south-west of the moor it is now common around Burrator. In 1957 P. J. Dare located about thirty singing males in the Postbridge area, and in 1964 a singing male was seen at North Teign Farm, at almost 1,500 ft. On west Dartmoor breeding was proved at Grenofen near Tavistock in 1965. In the same year about fifteen pairs bred at Yarner Wood, a regular breeding site on south-east Dartmoor.

On Lundy it occurs in small numbers as a regular spring and autumn passage migrant, during April and May, and August to early October, with maxima of fourteen on 17 April 1966 and thirteen on 22 September 1962.

BLACK REDSTART *Phoenicurus ochruros*
Passage migrant and winter visitor, has bred

The status of the Black Redstart in Devon does not appear to have changed appreciably since D & M's time. Then, as now, it occurred as a regular winter visitor and passage migrant from October to March, sometimes in late September and early April, more commonly in autumn than spring and principally on the north and south coasts, particularly the latter. Inland occurrences are much more infrequent, but there are records from many places including a number of scattered localities on Dartmoor.

The main passage occurs in October and November, with a smaller spring passage in March and wintering birds present during the intervening months. Most reports refer to ones and twos, the most seen together being a party of six at Lympstone in October 1940.

This species became more common during the 1940s, with 104 records for the mainland in the winter of 1948-9 and 169 in 1949-50, compared with a total of about 130 for the 9 years 1959-67.

Successful breeding by one pair was recorded at Burlescombe in 1942 and at Torquay in 1949; it may have occurred at Plymouth in 1941 but was not proved. A single male was present at Plymouth from November 1948 until November 1951, but did not breed.

Small numbers are recorded annually on Lundy in spring and autumn, and some winter. The highest daily counts were ten on 25 March 1949, twelve on 24, and nine on 25 October 1953.

NIGHTINGALE *Luscinia megarhynchos*
Summer visitor, breeds

Following an increase which culminated in over sixty singing males in 1932, and an apparent extension of its range westwards, the Nightingale has since decreased in numbers almost to the position it held during the latter part of the last century.

A scarce breeding bird in the nineteenth century, it was stated by D & M to nest only occasionally, principally in east Devon. Subsequently in the *VCH* it was reported to be breeding regularly in the Teign valley, and D'Urban's MS contains many records for the area from the Dorset border to the Teign, for the years 1900-1924.

That it certainly increased during this period and up to the 1930s is shown by the results of a survey which gave the annual number of singing males as ranging from forty-one to sixty-one in the years 1930-35, mainly between Axminster and Chudleigh Knighton, and including thirty in the Kingsteignton district in 1931. E. W. Hendy in 1936 considered it was spreading westwards and quoted records for a number of localities west of the Teign, including Hatherleigh, Northam, Instow, Eggesford, and Salcombe.

Although during the 1940s the numbers were fairly well maintained, they have decreased in all areas since about 1950, and vanished entirely from several, including Woodbury Common. Their stronghold in the lower Teign valley, which still held twenty-two singing males in 1952, produced only five in 1966. Several pairs still breed at long established sites in the Exeter district. Sporadic breeding further west included pairs at Ashburton, Stoke Gabriel, and the Plym valley during the 1950s.

Outside its main area, where it normally arrives during the second half of April, it is rarely encountered, but has very occasionally been recorded at Slapton in both spring and autumn. On Lundy it has been recorded on only five occasions since 1947 : four singles during the month of August and a single male in April 1963.

BLUETHROAT *Cyanosylvia svecica*

Scarce passage migrant

D & M cite two occurrences of the Bluethroat, which they regarded as a very rare, accidental visitor. The first relates to one shot near Whimple in September 1852, and the second, an unsatisfactory record, concerns one seen near Exeter in about 1869. Both are listed as the Red-spotted form (*C.s. svecica*).

There is no further record until one was seen on Lundy on 14 September and another on 20 and 22 October 1949, neither of which was assigned to a particular race. A first winter male, also of indeterminate race, was trapped and ringed on Lundy on 19 September 1956 and one, possibly the same, was seen two days later.

As a result of mist-netting in the reed beds at Slapton Ley during the 1960s, no less than nine examples, eight indeterminate and one showing the red spot, were caught and ringed in the years 1961-7. The first of these was trapped on 17 September 1961, four between

23 August and 3 September 1964, one on 17 September 1965, one on 11 and another on 21 September 1966, and the last on 18 September 1967. Thus, after a lapse of almost a century with no records for Devon, it appears that this extremely skulking bird has been passing through Slapton regularly on autumn passage, completely undetected, and but for mist-netting would probably remain so.

The individual ringed at Slapton Ley as a first year male on 21 September 1966 was retrapped in exactly the same spot on 14 September 1968, being then identified as an adult male Red-spotted Bluethroat. This is a most remarkable recovery of a passage migrant.

The only two spring records both refer to examples of the White-spotted race (*C.s. cyanecula*), one of which was closely watched by three observers at Bovisand near Plymouth on 29 March 1965, and the other was killed by a cat at Kingsbridge on 29 March 1968. The only inland record of this species during the present century relates to a bird seen at Holcombe Rogus on 14 September 1967.

ROBIN   *Erithacus rubecula*

Resident, breeds

The Robin, a resident and largely a sedentary species, is everywhere abundant and has a very wide breeding distribution. It is absent only from moorland with insufficient cover, but breeds plentifully in the conifer plantations on Dartmoor.

Although much reduced in numbers by the severe winter of 1963, it has since recovered and shows no appreciable change of status during the present century.

Autumn immigration is sometimes noted in coastal areas, eg twenty-two at Prawle Point, thirty at Wembury, and forty-five at Plymouth in September 1965. A bird recovered at Dartmouth in March 1960 had been ringed in the Gulf of Bothnia in September 1959.

On Lundy, where it is present throughout the year, from three to ten pairs breed almost annually, but none nested in 1963; breeding was again recorded in 1965 and 1966. A light passage is detected in some seasons, with peaks of thirty in October of 1944 and 1951, but movement is not recorded every year. Two birds trapped on 31 March 1965 were assigned to the Continental race (*E.r. rubecula*).

GRASSHOPPER WARBLER *Locustella naevia*

Summer visitor, breeds

A secretive bird, inhabiting rank undergrowth in marshes, heaths, bogs, and particularly in young conifer plantations, the Grasshopper Warbler has increased as a breeding species during the present century, and especially during the past decade. D & M considered it a very local species and quoted breeding records for only a few localities, including Dartmoor and Slapton.

On its arrival in mid-April to early May it is regularly recorded at many coastal places, where the birds stay for a while before moving on. At least twenty were observed at Hartland Point on 14 April 1959.

Omitting earlier records, definite proof of breeding since 1947, by actually finding the nest or young, has been obtained at Shebbear 1947, Sourton 1950, Torrington Common 1957, Gittisham Common 1959-60, Chudleigh Knighton 1964, and Wembury 1967. Other known breeding localities include Fernworthy, Bellever and Soussons Plantations, Holsworthy, Halwill, Plym, Hartland, Haldon and Ashclyst Forests, Woodbury Common, Slapton, Braunton Burrows, and Countess Wear. Seven pairs were reported breeding on Torrington Common in 1963; twelve pairs were noted in the Woolacombe/Morte Point area in June 1967, and the breeding population of the Postbridge area, including the plantations, for the years 1956-67 is given as from ten to twenty-five pairs. Breeding is presumed to have occurred at many other localities.

At Slapton, where fifty were ringed 1964-7, mostly in the autumn, a marked passage occurs during August and September.

This warbler occurs annually on spring and autumn passage on Lundy, usually in small numbers, but peaks of twenty-five were recorded on 23 September 1965 and nineteen on 16 April 1966.

MOUSTACHED WARBLER *Lusciniola melanopogon*

Vagrant

This Mediterranean species, which has been reliably recorded in the British Isles on only about six occasions, has occurred once in Devon when a single bird was observed by Michael Jones on Lundy on

2 May 1959. It was seen for about ten minutes in the garden at Millcombe, in company with Sedge Warblers, often in the same field of view. The account in the *Lundy Report* for 1959-60 emphasises the marked differences between this and the commoner species with which it was compared.

GREAT REED WARBLER   *Acrocephalus arundinaceus*

Vagrant

Although widely distributed on the Continent, the Great Reed Warbler is a rare vagrant to Britain and has been recorded only once in Devon. A single bird was observed by F. R. Smith at Slapton Ley on 7 August 1961 and reported in detail in the *Devon Report* for that year. Being familiar with the species in France and Spain, Mr Smith immediately recognised it as it perched quite openly near the bridge and, to quote his own words, was 'larger than a Reed Bunting; its shape and appearance was that of a giant Reed Warbler'. This occurrence is noted in *BB* 55:577.

REED WARBLER   *Acrocephalus scirpaceus*

Summer visitor, breeds

The Reed Warbler, a summer resident and passage migrant, has considerably extended its range in Devon during the present century. The only breeding localities known to D & M were Thurlestone, Milton and Slapton Leys, breeding being first recorded at Slapton by J. H. Gurney in 1871. The population there in 1956 was estimated as 200 pairs. Singing males numbered 146 in June 1963 and 120 in June 1964, indicating a decrease which was again noted in 1965-6.

Loyd stated in 1929 that it had bred on the River Otter, but there is no further detailed record of this until 1945, when nesting was again reported. Similarly, breeding was first reported on the Axe estuary in 1951, but probably commenced earlier. On the Exe estuary breeding was first proved in 1947, following which the species quickly spread to other reed beds on the estuary. In 1958 Reed Warblers were observed at Blaxton Marsh on the Tavy, where breeding was finally proved in 1964. In 1961 a singing male was reported at Chivenor on the Taw estuary, followed by the recording of two resident pairs at Horsey Pond in 1965, and breeding on the River Dart in 1967. Other localities include the Teign estuary and probably Beesands Ley.

Spring arrivals are usually recorded during the second half of April and early May, and the last departures in early October, while the main autumn passage is from late August to late September. At Slapton 1,454 birds were ringed in the years 1961-7. One of the more spectacular recoveries was a juvenile ringed there on 1 August 1965 and recovered at Sidi Ifni, Spanish West Africa, on 15 September 1965.

It occurs as a fairly regular but uncommon passage migrant on Lundy, where under thirty examples, all singles, were recorded from 1947-66.

MARSH WARBLER *Acrocephalus palustris*
Vagrant, breeds sporadically

Although this scarce summer visitor occurs in the neighbouring counties of Somerset and Dorset as an extremely local and rare breeding species, it is hardly more than a vagrant to Devon, despite having been proved to nest on one or two occasions, and suspected of doing so on a very few others. D & M, who did not include it in their work, considered that it very probably visited Devon.

The first record relates to a pair observed by W. Walmesley White on the Otter marshes for six days from 26 May 1925 and reported in *BB* 21:62. The next and most satisfactory record refers to a pair which bred at Slapton Ley in 1928 and of which a photographic record was obtained by A. M. C. Nicholl, who communicated the occurrence in the Society's *First Report*. The third record, which cannot be regarded as proven, refers to an observation by S. D. Gibbard on 22 June 1938 of a pair collecting food on the canal at Tiverton, and carrying it to a presumed nest on the other side of the water. No evidence of identification is given in the *Devon Report* for 1938, in which the event is recorded. The same journal for 1954 states that F. H. Lancum located a nest in the south of the county, and on two occasions watched the young being fed. At this stage it is no longer necessary to suppress the locality which, surprisingly, was Burrator reservoir. Brief notes of a pair observed in a suitable area on Braunton Burrows by J. Cadbury on 29 June 1957, and identified by song, are given in the *Devon Report* for that year, while fuller details of a pair observed in the same locality on 3 and 4 June 1960 are included in the 1960 *Report*.

Two occurrences in 1962 refer to one identified by song at Tamar Lake on 1 September, reported by S. V. Benson, and, the first record for Lundy, a single bird which was trapped and ringed on 2 September by J. Coleman Cooke and P. Griffiths, and listed in the *Lundy Report*. Another was recorded there from 12 to 16 October 1967.

SEDGE WARBLER   *Acrocephalus schoenobaenus*
Summer visitor, breeds

The Sedge Warbler, a summer resident, has a local breeding distribution, being mainly confined to marshy ground, with rank vegetation, along the lower reaches of the rivers. Its principal breeding areas, at some of which it is fairly common, are the estuarine marshes of the Axe, Otter, and Exe, and Slapton Ley in the south, and, in the north, Northam and Braunton Burrows and Braunton Marshes.

Small numbers also nest regularly on the Torridge at Torrington, on the canal at Sampford Peverell, at Tamar Lake, at Totnes and Sharpham on the Dart, on the Creedy, Culm, Clyst, Erme, Avon, and Plym, and probably a number of other localities. A population of thirty pairs was reported on the Axe estuary in 1962, and twenty pairs at Slapton in 1963-4.

In the Postbridge area of Dartmoor singing males have been seen in several localities since 1955, but although breeding was suspected in a few cases it has not been proved.

The first arrivals are usually noted in mid-April, and some birds are present until late September or early October. A considerable autumn passage occurs during August/September at Slapton, where more of this species are ringed than any other warbler; the total for 1961-7 was 1,933, compared with 481 for Lundy for the years 1947-66.

Except for sporadic breeding in 1934-5, the Sedge Warbler is only a regular passage migrant on Lundy, mainly in mid-April to early May and August/September, with maximum numbers of sixty on 6 May 1953 and 19 May 1962, and an exceptional autumn peak of 130 on 30 July 1965, the numbers fluctuating from year to year.

There is no evidence of any marked change of status in Devon since D & M's time.

AQUATIC WARBLER   *Acrocephalus paludicola*
Scarce passage migrant

The inclusion of the Aquatic Warbler in the Devon avifauna is another direct result of the work of the bird observatories. Unknown in the county in D & M's time, this species, which so closely resembles the Sedge Warbler, has now been found to occur almost every year, if not annually, as a scarce autumn passage migrant on the south coast and to a lesser extent on Lundy.

Most of the records are from Slapton Ley where eleven birds were trapped and ringed during the months of August and September, including one seen in early November, in the years from 1960 to 1968, principally due to the untiring work of F. R. Smith, amongst others. All the Slapton birds were mist-netted in the dense reed beds through which they would otherwise have continued to pass unnoticed. There are also two sight records from Dawlish Warren, one on 21 September 1956 and the other on 11 September 1964, all of which are duly listed in the *Devon Reports* for these years.

The five Lundy occurrences of this species comprise four sight-records—two on 15 September 1949, one on 31 August 1956, one on 3 May 1963, and one trapped on 13 September 1963. The May record is unusual as being the only recorded occurrence on spring migration, but it is not included with the 1963 records of the Aquatic Warbler in *BB* 57 : 273.

MELODIOUS WARBLER   *Hippolais polyglotta*
Scarce passage migrant

An irregular straggler on autumn migration, the Melodious Warbler is yet another species which has been reliably added to the Devon list by the painstaking work of the Lundy and Slapton Bird Observatories. Although this species was not mentioned by D & M, D'Urban subsequently included it in the *VCH* on the strength of a number of birds, supposed to be of this species, which were observed by the Rev M. A. Mathew at Ware near Lyme Regis in May 1897 and again in the same month of 1898. These records, however, together with that of a Melodious Warbler reported at Budleigh Salterton in May 1921, were rejected by the editors of *The Handbook* on the grounds that the birds were most probably wrongly identified.

The first fully substantiated record for the county is of a single bird which was trapped on Lundy on 30 July 1951, and whose description is given in great detail on pp 19-20 of the *Lundy Report* for 1951. A second bird was trapped on the island on 31 August 1954, within a few days of others being taken at Bardsey and Portland Bill, as related in *BB* 48 : 284. A third bird was trapped on 10 September 1958, and is recorded in the *Lundy Report* for that year, which notes that of some seventeen Melodious Warblers reported in Britain in 1958, thirteen occurred at the Irish Sea Observatories. This species was again reported in 1962 and 1963, when single birds were trapped and ringed on 15 September and 11 September respectively, as stated in the *Lundy Reports* for 1962 and 1963, and *BB* 56 : 404 (the 1962 record). There are also several sight-records for the island : three during September and October 1959; one in September 1962; one or more in September 1963; and two or three in August 1966; but most of these are indeterminate between this species and the very similar Icterine Warbler. These two species, although readily identified by their songs, are not easily separated in the field, away from their breeding quarters.

For the mainland, there is one fully authenticated record of a bird trapped and ringed at Slapton on 27 August 1964, and three reliable sight-records of this species during July and August 1956, when single birds were very closely observed at three different localities, viz Exmouth early July to 13 August (A. J. Ingham), Exeter 30 July (W. N. Bolderston), and Langtree between 21 July and 22 August (M. J. McVail). The last bird was also seen by F. R. Smith, who confirmed the identification, recorded in the *Devon Report* for 1956. Single birds at Dawlish Warren on 1 August 1965 and East Budleigh on 27 July 1966 were recorded as either Melodious or Icterine Warblers.

ICTERINE WARBLER   *Hippolais icterina*

Scarce passage migrant

This species, like the preceding, has been recorded in Devon on a number of occasions during recent years, and was in all probability previously overlooked, as it has now been found to occur irregularly as an autumn passage migrant through Lundy and less frequently on the mainland. The Icterine Warbler was not included by D & M, although they did mention in brackets a doubtful undated record from near Plymouth.

The first authentic record of this species refers to a bird trapped on Lundy on 28 August 1949 and examined in the hand. Another individual was seen on the island on four dates between 16 and 26 October of the same year. The *Devon Report* for 1949 also includes a sight-record of one observed beside the Otter estuary on 16 July, which should probably be regarded as indeterminate. Four Icterine Warblers, three of which were trapped and ringed, occurred on Lundy between 9 and 20 September 1951, while a bird which was either this species or a Melodious Warbler was seen on the island on three dates between 11 and 17 September 1954. There was a further occurrence in the following year, a single bird being caught on Lundy on 25 August 1955. Three or four more birds seen there between 10 and 12 September, three on 11, are listed as Icterine Warblers in the *Lundy Report* for 1959-60, as is one observed there on 11 September 1966, which is recorded in the 1965-6 *Report*.

In 1956 a bird seen at Prawle Point on 14 September by J. R. Brock, and carefully described, was identified as this species, while another observed at the same place by P. J. Dare, R. M. Curber, and R. Godsell, on 13 and 14 October 1962, and accurately described in the *Devon Report* for 1962, was accepted by the Rarities Committee as an Icterine Warbler and published in *BB* 56 : 404. As already stated in the account of the Melodious Warbler, there are a number of indeterminate records, and to those quoted should be added the examples recorded in the *Lundy Report* for 1961, relating to single birds watched on the island on four dates during August and September of that year.

BLACKCAP *Sylvia atricapilla*

Summer visitor, breeds

This very attractive warbler is widely distributed in suitable habitats, occurring quite commonly in open woodland and copses with thick undergrowth of brambles etc, and in large gardens and shrubberies. It is regular and in places common in conifer plantations, eg Plym Forest, but is absent from open country, including most of Dartmoor, except the afforested areas, where it is slowly spreading and

P

increasing, and has recently been recorded up to 1,500 ft. It is everywhere commoner than the Garden Warbler.

Small numbers of Blackcaps winter annually (over twenty individuals in 1961), and are then frequently recorded at bird tables.

Spring arrivals occur mainly during April; some March birds may have wintered. Autumn passage is recorded from July to late September at Slapton, where 570 were ringed 1961-7, compared with 218 Garden Warblers in the same period, and with only 213 Blackcaps ringed on Lundy 1947-66. One ringed at Slapton 14 September 1966 was recovered in Morocco 14 October 1966.

On Lundy, where it has not been proved to breed, the Blackcap occurs regularly in small numbers on spring and autumn passage, mostly in late April and early May, and from mid-September to the end of October, sometimes November. A larger than usual spring peak of twenty-one occurred on 16 April 1966; the largest number recorded in autumn was twenty on 23 September 1965.

BARRED WARBLER   *Sylvia nisoria*

Vagrant

D & M knew of no Devon records of this central European drift migrant, which occurs fairly regularly on autumn passage, chiefly on the east side of Britain.

Since the establishment of the Observatory on Lundy, the Barred Warbler has been reported nine times, all in September and October, indicating that the species does occur in the county as a vagrant, if not as a scarce and irregular passage migrant.

The Lundy records comprise a single bird on 10 October 1949, one from 23 September to 10 October 1960, with two on 28 September, one on 3 October 1966, and four between 18 September and 22 October 1966. The only other Devon record concerns a single bird observed by E. Griffiths at Wembury on the south coast on 12 October 1966.

GARDEN WARBLER   *Sylvia borin*

Summer visitor, breeds

Although widely distributed in suitable woodlands throughout the whole of Devon, the Garden Warbler is everywhere less common than the Blackcap. Like the latter, it also breeds regularly in coniferous woods and on Dartmoor has colonised the extensive plantations at Bellever, Soussons, and Fernworthy. It occurs, too, in the heavily wooded river valleys and on Exmoor is regularly reported from the Brendon area.

In spring it arrives later than the Blackcap, appearing in late April and the first week of May. A small autumn passage is noted between July and about mid-October at Slapton on the south coast. A first winter bird ringed there on 17 September 1963 was recovered in Morocco in September 1966.

Of the very small numbers that regularly pass through Lundy in spring and autumn, sixty-two were ringed in the years 1947-66. In most years fewer than ten are recorded in any one day, but a peak of thirty occurred on 23 September 1965. It is recorded in the *Devon Report* for 1934 that a pair nested on the island in that year; there is no previous or subsequent breeding record.

In the absence of any numerical data in D & M's account of this species, it is difficult to say whether there has been any significant change of status, but there is no evidence to suggest it.

WHITETHROAT   *Sylvia communis*
Summer visitor, breeds

The Whitethroat, an abundant summer resident, has a very wide breeding distribution. It occurs throughout the county in hedgerows and in rough ground with bushes, brambles, and gorse etc from the coastal cliffs to well up on Dartmoor and Exmoor. On the former, L. A. Harvey recorded it at approximately 1,300 ft at Black Tor Copse in June 1949, and P. J. Dare describes it as fairly common in the Postbridge district, especially in the partly cultivated areas.

Arriving during the first three weeks of April, it is present until about mid-September, sometimes later, with a marked autumn passage at Slapton from July to September. On Lundy, where from one to three pairs breed most years, eight pairs in 1952, it is abundant on both spring and autumn passage. The former occurs mainly from late April until early May and the latter from late July to about mid-October. Peak numbers recorded since 1947 include over 400 on 5 May 1953, an exceptional movement of at least 1,000 on 16 September 1958, and 900 on 8 September 1959; these, however, are greatly in excess of the usual peaks of up to about 150.

LESSER WHITETHROAT *Sylvia curruca*
Summer visitor, breeds

Although the Lesser Whitethroat has quite definitely increased in Devon during the present century, it is a scarce though probably regular breeding species, and the majority of singing males reported during April and May move elsewhere.

D & M, who never actually saw one in the county and considered it very rare, were able to cite only two definite occurrences. It is recorded in the *VCH*, however, as having nested at Tiverton (in 1886), Dawlish, and in the South Hams, and A. H. Rousham reported nesting at Exeter around 1907. Loyd found several breeding pairs at Beer in 1913, and W. Walmesley White stated in 1929 that two or three pairs had nested annually around Budleigh Salterton since 1915.

Breeding was reported at Staverton 1931, Plymouth 1937, Budleigh Salterton 1938-9, Lympstone (two pairs) 1939, Chudleigh Knighton 1945, Plym Valley 1947, Dalwood 1949, Farringdon 1954, Crediton 1955, Woodbury and Culm Davy 1957, Lympstone 1960, Ashburton 1960-61, Sidmouth (three pairs) 1961, Woodbury 1962, Budleigh Salterton 1963, Axe estuary (two pairs) 1964, Staverton 1966, and Lustleigh 1967. All these localities are in east and south Devon. Breeding at Barnstaple and Braunton in the 1930-40s is recorded in the *Ilfracombe Fauna & Flora*. The presence of singing males during June, at a number of other localities, suggests that breeding may have occurred on at least some of these occasions, especially at Crediton, where it has been consistently reported since 1956.

The first spring arrivals are recorded in the latter half of April, and a small autumn passage through Slapton is noted in late August and September. Singles occur most years on Lundy during April/May and August to October and occasionally November; the most recorded was four on 11-12 May 1965.

SARDINIAN WARBLER *Sylvia melanocephala*
Vagrant

An adult male Sardinian Warbler, the first accepted record of this

Mediterranean species for Britain, was trapped on Lundy on 10 May 1955. After being critically examined and photographed by Barbara Whitaker, then Warden of Lundy, it was released and soon disappeared into thick cover. A detailed account of this occurrence is given by Barbara Whitaker in *BB* 48 : 515, where it is stated to be the second British record; the first has since been rejected as one of the 'Hastings Rarities', *BB* 55 : 283-384.

A bird believed to be of this species was seen by D'Urban in his garden at Exmouth, together with a male Blackcap, on 16 April 1890 and is included in his *The Birds of Devon* under the name of the Black-headed Warbler. This record, however, was not fully accepted by *The Handbook*, in which it is qualified as possibly of this species.

DARTFORD WARBLER *Sylvia undata*
Resident, breeds

D & M feared that this delightful warbler had become extinct in Devon, as they knew of no occurrence since 1877, whereas it had formerly been resident in several parts of the county. However, their *Supplement* records that a few were seen at Kingsbridge in the autumns of 1893 and 1894. In the *VCH* D'Urban remarks that its nest had not been found in Devon since 1806 when Colonel Montagu discovered a small breeding colony near Kingsbridge and described the hitherto unknown nest and eggs.

There is no further published record until the Third *Devon Report* mentions a male seen by Colonel R. M. Byne on Woodbury Common in March 1924. In September 1939 H. G. Alexander found a small colony on Colaton Raleigh Common, and W. Walmesley White located three pairs and found a nest there in May 1940, and recorded breeding at two sites in 1943.

The *Devon Reports* confirm that breeding occurred in East Devon from 1946 to 1954, during which period it seemed that this very vulnerable species was becoming well established, following the discovery of a further colony in the south-east and the recording of odd birds at two or three other localities, including Little Haldon, Fernworthy, and elsewhere.

Although breeding was not proved in the years 1955 to 1966, singles, twos, and threes were observed most years in two or three localities, including two migrants on the south-west coast, one at

Portlemouth and the other at Gammon Head, both on 13 October 1960.

In 1967 one pair bred at a locality which, for the sake of the birds, was undisclosed.

Further evidence of migration of this normally resident warbler is provided by two records from Lundy, where singles were observed on 28 October 1951 and 26-28 March 1963.

RUFOUS WARBLER   *Cercotrichas galactotes*

Vagrant

Devon can claim three occurrences of this rare straggler from the Mediterranean. Two records, both of birds obtained, are listed by D & M, the first relating to one that was shot near Start Point on 25 September 1859, and the second to a bird shot at Slapton on 12 October 1876. The most recent occurrence, and the only one for the present century, concerns a single bird which also occurred on the south Devon coast at Gammon Head near Prawle Point on 20 October 1959 where it was seen by R. C. Stone and E. C. Still, whose detailed description of the bird appeared in the *Devon Report* for 1959 and *BB* 53 : 225.

WILLOW WARBLER   *Phylloscopus trochilis*

Summer visitor and passage migrant, breeds

An abundant and very widely distributed summer visitor, the Willow Warbler breeds throughout the county, in open woodland, young conifer plantations, copses, and bushy ground, and greatly outnumbers the Chiffchaff.

It is plentiful on Dartmoor wherever there is sufficient cover and has been recorded breeding at 1,500 ft. In the 15 square miles around Postbridge the population is given by Dare and Hamilton as 250-500 pairs. It is described in all local lists as common or abundant, and is recorded as breeding in almost every 10 km square. There is no evidence of any change of status.

An early migrant, it arrives from the last week of March to about mid-April, and autumn passage is recorded on the coasts from July to September, with the main numbers in August.

On Lundy, where it breeds sporadically, it is abundant on spring and autumn migration, the former mainly in April and early May and the latter chiefly from late July to September. Peaks of 800 were recorded on 28 April 1959 and 600 on 16 September 1958, but these are considerably in excess of the normal peaks of 100-200. In the twenty years 1947-66 a total of 2,145 Willow Warblers were ringed on Lundy compared with 622 Chiffchaffs, whereas at Slapton from 1961-7 only 386 were ringed compared with 1,066 Chiffchaffs. Breeding on Lundy was recorded in most of the years 1927-44 (four pairs in 1930), probably 1947, one or two 1949, and probably 1954.

GREENISH WARBLER *Phylloscopus trochiloides*

Vagrant

An example of this rare leaf warbler was seen on Lundy on 2 November 1958 and was mist-netted on the following day when its identification was confirmed after a detailed examination in the hand by W. B. Workman. What was thought to be a second bird of this species was seen on the island on the next day. The first occurrence, recorded in the *Lundy Report* for 1958, constitutes the only record for Devon of a north-central European and Asiatic species which has been extending its range westwards in recent years.

This record was accepted by the Rarities Committee and published in *BB* 53 : 412.

CHIFFCHAFF *Phylloscopus collybita*

Summer visitor and passage migrant, breeds

The Chiffchaff, a common summer visitor, breeds in woods, copses, shrubberies, and suitable gardens throughout the county, except on high ground. Although common in the fringe woodlands, it is scarce high up on Dartmoor, where a few pairs breed only in sheltered localities. Preliminary work on the BTO Atlas shows it to breed in practically every 10 km square.

An early migrant, it arrives usually from mid-March onwards and is present until August or September. A marked autumn passage occurs on the south coast between July and October, and at Slapton Observatory over three times as many Chiffchaffs as Willow Warblers have been ringed. A few wintering birds are recorded in scattered localities every year. There is no evidence of any change of status this century.

It occurs on Lundy as a regular and common passage migrant, mainly in April and from August to October, with occasional records in most other months, but generally in smaller numbers than the Willow Warbler. One trapped on 15 November 1952 was stated to be of the Scandinavian race, *P.c. abietinus*, as were six seen in November 1956. Some others in recent years were considered to be of the Siberian race, *P.c. tristis*. Proof of breeding on Lundy was obtained with one pair in 1965.

WOOD WARBLER   *Phylloscopus sibilatrix*

Summer visitor, breeds

A summer resident, the Wood Warbler breeds in open, mature deciduous woodland in all parts of the county, from east to west and from the coast to well up on Dartmoor, its distribution being dependent on suitable woodland. Although local, it is common in some areas, particularly in the oakwoods of the Dartmoor and Exmoor river valleys, and in many other woods with mature beech trees.

On Dartmoor singing males have been recorded at 1,500 ft at Princetown, and at Wistman's Wood, Black Tor Copse, Pile's Copse, and Fernworthy, while at Bellever breeding was confirmed in 1961. In 1957 eighteen singing males were reported at Yarner and the same number at Plymbridge in 1965, but this may be in excess of the number of breeding pairs. There is no indication of any decrease in Devon, and it may well have increased slightly.

Spring arrivals are usually first seen towards the end of April, sometimes early May. Just occasionally an autumn migrant is noted on the south coast during September, but, unlike the Chiffchaff and Willow Warbler, there is no marked spring or autumn passage; no example has so far been ringed at Slapton, and only four on Lundy, in the twenty years to 1966.

It occurs irregularly on Lundy, a few, mostly singles, having been recorded in about eight springs and seven autumns during the years 1947-66. The most recorded there in one day was six on 20 August 1959.

BONELLI'S WARBLER *Phylloscopus bonelli*
Vagrant

The second example of this south European leaf warbler to be recorded in Britain was trapped on Lundy on 1 September 1954. A full account of the occurrence is given by Barbara Whitaker on p 23 of the *Lundy Report* for 1954. A second occurrence on the island, involving two birds on 26 August 1960, is recorded in the *Lundy Report* for 1959-60, but unfortunately gives no details other than the number and date. This species is not, of course, mentioned by D & M.

ARCTIC WARBLER *Phylloscopus borealis*
Vagrant

An example of this rare leaf warbler, the first and only record for Devon, was mist-netted on Lundy on 6 September 1959 by W. B. Workman. Examination in the hand revealed the very short first primary which characterises this species. A brief description of the bird is given in the *Lundy Report* for 1959-60, and the record was accepted by the Rarities Committee, who published it in *BB* 53:425.

YELLOW-BROWED WARBLER *Phylloscopus inornatus*
Vagrant

There is no record of this Asiatic leaf warbler in D & M, but it has since been reported on at least sixteen occasions from 1932 onwards, seven occurrences being in the months of March to May and the rest in October and November. Whether certain of the records would now be passed by the Rarities Committee is open to doubt, but there is no question of the accuracy of a number of them. Of these records nine or more refer to Lundy where the first was reported by F. W. Gade on 26 April 1944 (BL 90); another was seen on 6 and 7 October 1949; one was seen on 15 October 1950 and, probably the same individual, was trapped and examined in the hand on 22 October; one was observed from 9 to 12 November 1955; one was seen by M. R. Jones on 4 and 5 October and, probably the same bird, on 6 October 1962, and was accepted by the Rarities Committee and published in *BB* 56 : 405; four were trapped and others were seen between 11 to 27 October 1967.

The eight records of this very small warbler on the mainland comprise a bird seen at Stoke Woods, Exeter, on 27 March 1932 (A. O. Rowden); one seen by W. L. Colyer at Sidmouth on 19 November 1941 (*BB* 35 : 181); one seen on the River Otter on 11 March 1950 by J. M. Reese, which is described in the *Devon Report* for 1950; and on 9 April 1952 F. R. Smith detected three beside the River Exe at Countess Wear, which he fully described in the *Devon Report* for 1952, which also lists one seen by N. Ross on 15 April of the same year at Muddiford near Barnstaple. A bird of this species closely watched by G. H. Gush on the scrub covered cliffs at Torquay on 31 May 1957 is well documented in the *Report* for that year. There is a sight-record of one at Burnt Halt near Tiverton on 7 and 8 April 1958, though it was not included in the first report of the Rarities Committee for that year. Lastly, one was closely observed by M. R. Edmonds at Slapton Ley on 8 October 1968.

GOLDCREST *Regulus regulus*

Resident and winter visitor, breeds

Resident, generally distributed, and common in D & M's time, the Goldcrest has steadily increased during the present century with the planting of conifers, and is one of the commonest breeding species in the state forests, as well as occurring throughout the county in other mixed woodlands, parks, and gardens containing conifers. Although the numbers are drastically reduced by severe winters, especially of the severity of 1962-3, the species quickly recovers and within 3-4 years is back to strength.

It is now abundant in the plantations at Burrator, Bellever, Soussons, and Fernworthy on Dartmoor where it occurs at levels up to 1,500 ft.

On Lundy it occurs regularly on spring and autumn passage, the former during March and April and the latter during August to late October. The maximum counts are fifty on 18 April 1962 and over forty in October 1959, but the normal figures are much less. Birds trapped in October 1952 were considered to be of the continental form *R.r. regulus*. Loyd records that one pair bred in 1922 and 1923, and the *Lundy Report* for 1952 states that a pair nested successfully in Millcombe.

FIRECREST   *Regulus ignicapillus*
Passage migrant and winter visitor

The minute Firecrest, regarded by D & M as a very rare bird, has been recorded in 18 of the 25 years from 1943 to 1967. During these years over eighty-seven birds were observed, usually singly, but sometimes two or three and, rarely, up to five together. The great majority occurred at places along the entire south coast, usually right on the coast, but sometimes a few miles inland. There are only about two records for the north coast, one from Combe Martin and the other from Barnstaple, while inland occurrences include Shobrooke, Wrangaton, Stover, Crediton, and Scorhill. Most of the birds are on autumn passage along the coast, but the December, January, and February records relate to wintering birds. Of the eighty-seven individuals recorded, about twenty-one occurred in November, nineteen in October, thirteen in December, eleven in March, nine in February, and the remainder in April, July, and September.

Years in which more than usual occurred were 1953, with about twelve records, 1955 with nine, and 1967 with at least twenty-four, including five birds at East Prawle on 22 October and two other records of three each.

In addition to the records for the mainland, the Firecrest occurs almost annually on Lundy, where at least twenty-five birds have been recorded between 1949 and 1966, mostly during the months of October and November, but including one in August. An early autumn record for the mainland relates to a bird trapped at Slapton on 13 July 1963.

SPOTTED FLYCATCHER   *Muscicapa striata*
Summer visitor, breeds

This quiet and unobtrusive bird is widely distributed throughout Devon, breeding fairly commonly in gardens, parks, farms, and the edges of woods, in all areas, including much of Dartmoor and Exmoor, wherever there are habitations.

A late migrant, it arrives on average in the second week of May and is present until about mid-September. Small flocks of migrants are recorded along the south coast from July to September. There is no evidence of any change of status since the last century.

Nesting on Lundy was first proved in 1956, when one pair bred successfully—followed by a pair in 1957, 1959, 1960, 1962, and 1963. The species occurs there regularly in varying numbers on spring and autumn passage, being sometimes scarce in spring but occasionally plentiful. A marked passage occurred in 1959, with a peak of 200 on 22 May; peaks of around 100 were recorded on 12 May 1960 and 9 May 1964. Autumn passage occurs mainly during August and September, with peaks of seventy-five on 22 August 1949, eighty on 13 August 1955, and 100 on 1 September 1958. Occasional stragglers have been recorded on the island during October.

PIED FLYCATCHER   *Ficedula hypoleuca*

Summer visitor, breeds

The Pied Flycatcher, regarded as a scarce bird in D & M's time, is now recorded regularly on spring and autumn passage and has become firmly established as a breeding bird on Dartmoor. D & M knew of no definite breeding record but thought that a pair nested near Barnstaple in 1859.

Sporadic breeding occurred during the first half of the present century : near Culmstock in 1907 and 1911, Templeton near Tiverton in 1911 (*BB* 5:134), and near Chagford 1949-51. Breeding in nest boxes erected by Dr Bruce Campbell at Yarner Wood has occurred annually since 1955, when the first pair nested successfully. The population there increased to six pairs by 1959, varied from two to six nests until 1965, and increased to nine pairs in 1966, while in 1967 eighteen pairs reared ninety-seven young, followed by a further increase in 1968, when twenty-six pairs reared 134 young.

A pair nested at Tawstock in 1956, Whitestone Wood in 1958, Wrangaton 1966-7, and Hollacombe near Winkleigh, Awliscombe, Dendles Wood, and Lustleigh in 1968. Singing males have been reported at several other localities including Glenthorne in 1960 and Merton in 1961.

Small passage movements are nowadays observed on both coasts, particularly the south, in April/May and August/September, with a maximum of eleven at Prawle Point on 9 September 1956.

Regular observation on Lundy since 1947 has shown the species to occur almost annually in spring, and annually and in greater numbers in autumn, with peaks of about forty on 6 September 1953 and sixty on 21 September 1966.

COLLARED FLYCATCHER   *Ficedula albicollis*
Vagrant

A rare straggler from central and south-east Europe, the Collared Flycatcher has been recorded once in Devon and up to the mid-1960s only about five times in the British Isles. The Devon record, reported in *BB* 42 : 292, refers to a single male which was observed by Mrs D. Wilson and D. ffrench-Blake close to Baggy Point, north Devon, on 20 April 1948. The characteristic bold white collar which separates this species from the otherwise similar Pied Flycatcher was immediately apparent to both observers.

RED-BREASTED FLYCATCHER   *Ficedula parva*
Scarce passage migrant

D & M remarked in a footnote that the Red-breasted Flycatcher might be expected to occur in Devon, having been obtained several times in Cornwall and the Scillies. When they wrote this, however, it had not been recorded in the county, and it was not until the Lundy Bird Observatory came into operation that this species was definitely recorded and subsequently found to pass through Lundy with some degree of regularity during the autumn.

The first record occurred on 20 October 1950, since when it has been recorded on the island in 9 subsequent years up to 1967, involving about seventeen birds, with no more than one in any year except for two in 1951, four in 1959, and four in 1967. Thirteen appeared in October, one in September, and three in November.

For the mainland of Devon there are only two satisfactory records, both of males and both in 1958, when one was seen near Ivybridge on 11 September by E. Honour, and the other at Mary Tavy on 27 September by A. R. Smith. Both birds are adequately described in the *Devon Report* for 1958. A doubtful occurrence, mentioned in *BB* 15:142, concerns two birds which were said to have frequented a garden at Exmouth during the summer of 1921.

DUNNOCK *Prunella modularis*
Resident, breeds

The unobtrusive Dunnock, or Hedge Sparrow, is an abundant breeding bird with a wide distribution. It is resident and plentiful in all areas except the highest parts of Dartmoor, but even there it occurs wherever there is sufficient cover, having been recorded at 1,540 ft at Assycombe Hill and 1,500 ft at Holming Beam, and being quite common in the more cultivated areas.

It is resident and breeds on Lundy where the population has been variously given as two pairs in 1922-3 (Loyd), twenty-three pairs in 1930 (Wynne-Edwards and Harrison), six in 1939 (Perry), and varying from two to at least twenty pairs between 1948 and 1967. As the species remains on Lundy throughout the year, the detection of any migratory movement is difficult, but a small movement has been suspected.

There appears to have been no change of status since 1830 when Dr E. Moore wrote that it was 'very common all the year'. The numbers are affected, however, by particularly severe winters such as 1962-3 but they recover and the species is again plentiful.

ALPINE ACCENTOR *Prunella collaris*
Vagrant

The Handbook admits four occurrences in Devon of the Alpine Accentor, a rare straggler from the mountain ranges of south and central Europe. These very old records, the last of which occurred over 100 years ago, relate to one bird killed on the cliffs at Teignmouth on 9 January 1844; another, taken in the same year at Berry Head, which is now in the Torquay Museum; and two obtained near Plymouth on 10 January 1859. There has been no further recorded occurrence of this species in the county.

MEADOW PIPIT   *Anthus pratensis*
Resident and winter visitor, breeds

As a breeding bird the Meadow Pipit is widely distributed throughout the county, occurring wherever there are commons, moorland, rough grasslands, and downs, both inland and along the entire coast. It is abundant on both Exmoor and Dartmoor where it breeds right up to the highest and most desolate areas, being often the only passerine species.

The moors are practically deserted from about October to March, when the species congregates in lowland and coastal districts, augmented by winter visitors. Migratory movements are noted most years, chiefly in September and October, when flocks are seen moving mainly westwards, along the south coast, though they have been recorded flying due south from the north coast—eg 1-5 October, a southerly movement from Lynton recorded daily, with a peak of 800 in one hour on 5 October 1961.

Considered the most numerous passerine species breeding on Lundy, the population was given as about 275 pairs in 1930, and 200 pairs in 1939. In 1962 it was recorded as about fifty pairs, but in 1965 as only about thirty birds. A marked spring and considerable autumn passage are noted, the numbers recorded varying greatly from year to year. Large numbers occurred in October 1954, with a peak of over 2,000 on 7 October, while in 1959 there was a movement of 1,000 or more on 22 September.

There is no indication in Devon of the decrease in this species noted in some other counties.

RICHARD'S PIPIT   *Anthus novaeseelandiae*
Vagrant

D & M record that eight examples of this Asiatic pipit were obtained, and another seen, in the Plymouth district between the years 1841 to 1878. Three others were shot on Braunton Burrows, one in December 1864, and the other two in January and December of 1872. The next occurrence is of two seen by W. L. Colyer on the Otter estuary on 10 February 1942. A pipit most probably of this species was observed on Braunton Burrows on 12 October 1961, while a further example occurred on the nearby Northam Burrows on 1 September 1963 and was recorded in *BB* 56:275. About eleven of these large pipits have been recorded on Lundy, all in the months of September and October in 1957, 1958, 1963, 1966, and 1967, details of which are given in the *Lundy Reports*. The most occurring together was three on 24 September 1966, *BB* 60:327. Whether the two birds seen at Exmouth on 6 February 1945 were correctly identified as this species cannot now be judged from the meagre evidence. A further six were recorded in the autumn of 1968.

TAWNY PIPIT   *Anthus campestris*

Vagrant

The Tawny Pipit, although having a wide distribution in Europe and Asia, is no more than a vagrant to Britain, and in D & M's day had not been recorded in the county. This wagtail-like pipit is another of those species which have been added to the Devon list through the work of the Lundy Field Society, four single examples having been recorded on the island during the past twenty years. The first occurred on 6 and 7 October 1950 and is adequately described on p 16 of the *Lundy Report* for 1950. What were considered to be two different individuals were seen on 19 and 29 September 1951, the latter bird being watched for ninety minutes in the field, as related on p 19 of the *Lundy Report* for that year. The fourth bird, which like the others was on autumn passage, was recorded on 2 November 1966.

TREE PIPIT   *Anthus trivialis*

Summer visitor, breeds

Occurring as a summer visitor in hilly rather than lowland areas, the Tree Pipit breeds annually in all suitable localities throughout the county, except for the extreme west, for which there are few records. It is widespread on Dartmoor wherever there are scattered trees, particularly on the wooded fringes of the south and south-east, and also regularly in young conifer plantations at Fernworthy, Bellever, Soussons, and Burrator, and has been recorded at Wistman's Wood. It breeds regularly on the heaths and commons such as Haldon, Woodbury, and Chudleigh Knighton, and occurs in many parts of north Devon, including the Torrington district, Braunton, and the north-western fringe of Exmoor. In the south-west it is regular in the Bickleigh area; it also occurs on rough ground in the South Hams, in agricultural country around Crediton, and in parts of East Devon.

Although nowhere abundant, it holds its own and there is no indication of any decrease during the present century. Spring arrivals are recorded from mid-April until early May, and autumn movement occur mainly from about late August until the end of September. The Tree Pipit is recorded annually as a passage migrant on Lundy and, though infrequent and sometimes absent in spring, is regular in autumn, when peaks of up to about twenty-five are noted during September. There is no breeding record for the island.

RED-THROATED PIPIT *Anthus cervinus*

Vagrant

There is one Devon occurrence of the Red-throated Pipit, a straggler from the tundras of northern Europe and Asia. The *Lundy Report* for 1959-60 briefly summarises the 1959 occurrence as follows: 'Two, almost certainly a pair, were in the tillage field on May 7th and still there on 8th, when one was caught and ringed, the first to be ringed in Britain. This species has not previously been recorded on Lundy'. For some unknown reason this record does not appear to have found its way into *BB* journal.

ROCK and WATER PIPITS *Anthus spinoletta*

(a) Resident, breeds
(b) Passage migrant

The Rock Pipit (*A.s. petrosus*) is a resident species which breeds commonly along the rocky shores of both the north and south coasts. Being an unobtrusive bird that is always present and one whose numbers hardly vary from year to year, it has rarely been counted. An indication of its breeding strength, however, is suggested by a record of four nests found in 150 yd of coast at Thatcher Point in 1958.

R

During the winter it is more widely dispersed and occurs along sandy shores and well inside estuaries, often some miles from the coast, as in the case of one observed on the Tavy in March 1956. Inland occurrences are most unusual, but there is a record of two, presumed migrants, at Tamar Lake on 7 April 1958, and two which were some miles up the River Otter at Colaton Raleigh on 8 April 1958. Of the few records of numbers, mention may be made of about thirty-five birds which were noted on the sandy shore at Bantham in December 1967.

Although evidence of some migration is provided by the recovery at Barnstaple in January 1963 of a bird ringed in Sweden, there appears to be no proof of the large flocks arriving in winter referred to by D & M. If it were so, it would most probably be apparent on Lundy where, in fact, there is no indication of any movement. The Rock Pipit is resident on Lundy throughout the year and the number of breeding pairs has been variously estimated or counted as fifteen in 1948, thirty-six in 1949, forty in 1951, and twenty in 1966.

The Water Pipit (*A.s. spinoletta*), of which D & M knew of only two examples, occurs as a rare winter visitor and a rather more frequent passage migrant, chiefly during April. The nineteen occurrences, all since 1956, involve about thirty-three birds of which twenty-one occurred in April, eleven in December to March, and one in September. Most records refer to singles, but nine in breeding plumage were identified at Hallsands on 12 April 1966 by S. C. Madge; up to three have been seen on Chelson Meadow, Plymouth during April; and single wintering birds have been noted there during recent years. All the records are from the south coast, and mostly from estuaries. Further observation will undoubtedly show the Water Pipit to be more frequent than formerly supposed.

PIED and WHITE WAGTAILS   *Motacilla alba*

(a) Resident and passage migrant, breeds
(b) Passage migrant, has bred

A common and widely distributed bird, the Pied Wagtail (*M.a. yarrellii*) breeds throughout the county, including much of Dartmoor and western Exmoor wherever there are farms, though most move to lower ground in winter.

Spring and autumn migrations are observed annually on both coasts; particularly the autumn passage, when flocks of 100-200 are recorded from August to October in many localities, roosting in reed beds, bracken, gorse, and even on moored boats. A peak of 600 was recorded at Slapton on 8 October 1955, and over 500 in many days during autumn 1963. Records of winter flocks include one of 400 roosting at Dawlish Warren in December 1958.

The Pied Wagtail breeds sporadically on Lundy and occurs regularly on spring and autumn passage. Breeding occurred in nine of the years 1947-66, with a maximum of five pairs in 1962.

The White Wagtail (*M.a. alba*) occurs regularly on spring and autumn passage along both coasts, occasionally inland and regularly on Lundy, but is identified chiefly in spring plumage. The numbers are usually under twenty, but records of larger flocks include seventy at Chivenor on 29 April 1939, sixty on Northam Burrows on 28 April 1954, and 150 at Westward Ho! on 21 April 1958. Several ringed in Iceland have been recovered in autumn on Lundy and at Slapton. Of an autumn movement of fifty on 7 September, seventy on 11 September, and 120 on 14 September 1961, through Lundy, most were of this race.

A pair of White Wagtails bred successfully on Lundy in 1956.

GREY WAGTAIL *Motacilla cinerea*
Resident, breeds

A mainly resident species, the Grey Wagtail shares with the Dipper the moorland streams and rivers, but is more widely distributed than the latter, as it breeds also on small lowland streams, close to habitations. On Dartmoor a pair bred successfully in 1965 at Teignhead Farm, situated at 1,450 ft. It is recorded as breeding in all parts of Devon.

Many of the moorland sites are deserted after the breeding season, when the birds move to lower ground and frequent the sides of estuaries, the seashore, and other waterside habitats. A small autumn passage is detected most years during September and October, when parties of up to about ten are noted in coastal areas, often with Pied Wagtails.

It occurs regularly on Lundy, in very small numbers on spring and autumn passage and occasionally during the winter, the records covering all months except December. The maximum number recorded in one day was twelve on 17 September 1953. There is no reliable record of its ever having nested on the island.

Except for a temporary but severe reduction in numbers caused by the hard winter of 1963, its status does not appear to have changed to any appreciable extent since D & M listed it as a common breeding bird at the end of last century. The records for 1966-7 indicate that it is again resident in all areas and has practically regained its former position.

YELLOW and BLUE-HEADED WAGTAILS *Motacilla flava*

(a) Summer visitor and passage migrant, breeds

(b) Passage migrant, has bred

A local breeding species and common passage migrant, the Yellow Wagtail has increased as a breeding bird during the present century. D & M stated that it rarely bred, and cited only two nesting records: Plymouth 1872 and Huish Marsh 1893. D'Urban's MS records that it bred on the Otter marshes in 1924, where it evidently continued to nest until about 1930. A. W. Mayo informed me that he found three nests on the Axe marshes in 1930, and a pair is recorded as breeding at Chudleigh Knighton in 1932.

The main breeding area is Exminster marshes where nesting was first proved in 1938, but probably occurred earlier. Breeding by a few pairs has continued at this site during subsequent years, the population gradually increasing to fifteen to twenty pairs in 1959-62 and around twelve pairs in 1963-5. Meanwhile, a pair bred at Wilmington in 1948, and on the Clyst marshes at Topsham in 1962 and 1965 and probably other years. Breeding on the Axe estuary was again reported, with three or four pairs, in the years 1963-6; and one pair bred at Braunton in 1960 and three pairs at Plymouth in 1966.

Spring and autumn passages are regularly observed on both coasts, the former from early April until early May and the latter from about mid-July to late September, occasionally October. Autumn flocks of up to thirty or forty are seen regularly at many coastal localities, particularly in the south, and sometimes inland, and roosting flocks of 100-200 are not infrequently reported at the reed beds on the Axe and Exe and at Slapton Ley, during August and September. One such roost on the Exe estuary in August 1949 contained over 500 birds.

The Yellow Wagtail is unusual on high ground, but occasional birds have been observed on Dartmoor during the autumn. It occurs on Lundy as a regular spring and autumn passage migrant, and a pair is recorded in the *Journal of Ecology* 20 : 374 as having nested there in 1930.

The Blue-headed Wagtail (*M.f. flava*) occurs in Devon as a scarce and irregular passage migrant, a few being recorded in coastal areas most years, usually with Yellows. It has several times been reported on Lundy. Loyd considered that a pair nested near Seaton in 1913, and breeding was recorded on Exminster marshes in 1959. Further examples, resembling other races, have been recorded from time to time, including a pair which nested successfully at Plymouth in 1965, the male of which had the characteristics of the Ashy-headed Wagtail (*M.f. cinereocapilla*).

WAXWING *Bombycilla garrulus*
Irregular winter visitor

The periodic irruptions of Waxwings from the forests of northern Europe rarely bring more than the occasional straggler so far to the south-west as Devon, and in many of the invasions none at all are recorded. Indeed, surprisingly few were listed by D & M, whose latest record referred to four birds in 1850.

For the present century there appear to be no records of any importance prior to 1928. From then until 1967 Waxwings were recorded in fourteen years, but in ten of them the total amounted to only fifteen singles. In addition, four or more were reported at Exeter from February to April 1958 and four at Brixham in January 1964.

The only large invasion to reach Devon during the present century occurred in the winter of 1965-6, when appreciable numbers were recorded during the last week of November, gradually decreasing during December, and with a few stragglers present until March 1966. The largest numbers occurred in south-east and south Devon, where flocks of up to thirty were reported in fourteen localities between Dawlish and Dartmouth, and flocks of up to twenty at thirteen localities between there and Plymouth. Singles and small parties of up to five (one of nine) occurred in east, mid, north, and north-west Devon.

All the recorded occurrences have been in the months from November to April, and the majority of birds have been reported from coastal areas. There is no fully substantiated record for Lundy.

GREAT GREY SHRIKE *Lanius excubitor*
Irregular winter visitor

About twenty-seven occurrences of this species during the nineteenth century were cited by D & M, who regarded it as an occasional winter visitor. The pattern is much the same at the present time, with over thirty records for this century, all but one referring to single birds, and all recorded during the months from October to April. The eight occurrences for April suggest a small passage during that month.

The occurrences since 1941 are fairly evenly spread over the whole county and include Hartland, Holsworthy, Exmoor, Braunton Burrows, east and west Dartmoor, Haldon, Payhembury, and a number for Woodbury Common, where the species has been recorded in eight years since 1956. Possibly because of the greater number of observers now, this shrike has been more frequently recorded in recent years, having been seen somewhere in the county in 15 of the past 20 years. The only instance relating to more than one bird was the occurrence of two on Woodbury Common on 27 February 1965.

There is no substantiated record for Lundy.

LESSER GREY SHRIKE *Lanius minor*
Vagrant

A very much rarer bird than the preceding species, the Lesser Grey Shrike has been reliably recorded in Devon on only two occasions. D & M quote two examples—an immature bird that was captured by a bird-catcher near Plymouth on 23 September 1876, and a probable Lesser Grey Shrike observed near Budleigh Salterton on 22 July 1894—but only the former, which was vouched for by J. Gatcombe, is a satisfactory record. The second fully authenticated occurrence relates to an adult which was seen on Lundy on 24 September 1961 by A. J. Vickery, who particularly noted the broad black band running from behind the eyes right across the forehead. This record, which appears in the *Lundy Report* for 1961, was accepted by the Rarities Committee and reported in *BB* 55:581. The *Lundy Reports* also include two probables, on 14 September 1958 and 26 May 1963, neither of which was accepted as being fully substantiated.

The only other record mentioned in the *Devon Reports* is that of a bird which alighted on the yacht *Provident* when 12-15 miles south-east of Start Point on 6 May 1952. The bird was thought to be of this species but cannot be claimed as a valid record.

WOODCHAT SHRIKE   *Lanius senator*
Scarce passage migrant

Some doubt attaches to most of the old records of the Woodchat Shrike, except for that of an immature female shot by E. A. S. Elliot near Bantham at the mouth of the Avon on 2 September 1892. The sight-record of an adult male observed near Lyme Regis on the eastern border of Devon on 22 June 1876 is included by D & M but claimed as a Dorset record by Blathwayt.

For the present century there is no mention of this very handsome shrike until 1939, when one was reported to have been seen on Halsinger Down near Braunton on the unlikely date of 12 March, but the record is not fully substantiated and lacks conviction. There is, however, an authentic recent record for the mainland of a male seen at Seaton Landslip by D. E. Paull on 1 June 1958, which was accepted by the Rarities Committee and is published in *BB* 53:171.

It was not until the start of regular observations on Lundy that this species was found to be a fairly regular migrant through the island, on spring and autumn passage. Since then single birds have been recorded there in ten of the years between 1949 and 1967. Of the fifteen records involved, six occurred during the month of May, three each in June and August, and one each in April, July, and September.

RED-BACKED SHRIKE   *Lanius collurio*

Scarce summer visitor, breeds

The Red-backed Shrike was regarded by D & M as 'a summer migrant, not very common, but frequently met with in the neighbourhood of the sea-coast both in the north and south of the county'. It is, unfortunately, a species which has been decreasing over most of its range in Britain during the present century, and particularly in the past 20 years or so. Of the main breeding localities in Devon in the early 1930s, Gittisham Common and the Ilfracombe area were deserted in the late 1930s, Saunton and Downend in the late 1940s, and Chudleigh Knighton, where up to three or four pairs bred, was deserted in the late 1950s. Meanwhile, these birds were slowly declining in numbers at their stronghold on Woodbury, Bicton, and East Budleigh Commons, where twelve pairs were recorded as breeding in 1947, seven in 1960, and three pairs in 1967.

Loyd, in *BSED*, stated that at least six pairs bred in the parish of Branscombe in 1915; the number was reduced to one pair in 1919, four pairs in 1922, and none in 1926. Sporadic breeding has taken place at a number of other localities, including Musbury in 1943, Haldon and Dawlish in 1945, Slapton in 1958 and 1959, and possibly near Start Point in 1961. There are very few records of occurrences on Dartmoor, but a pair was seen at Soussons Down on 14 July 1958.

This very attractive bird occurs as a scarce and irregular passage migrant on Lundy, where some sixteen individuals have been recorded, usually singly and mostly in September, in about 9 of the 20 years from 1947 to 1966. The species is said to have nested on Lundy in or about 1909, but there are no details. There appears to be no record of its ever having bred in Cornwall, while in Somerset, as in Devon, it has decreased very considerably, and it would appear to be only a matter of time before it is lost to the county as a breeding species. The reason for the decrease is not known and does not appear to be due to loss of habitat, despite the increasing amount of disturbance during recent years. An account of the decline and status of this species in Britain, by D. P. Peakall, was published in *Bird Study* 9 : 198-216.

STARLING   *Sturnus vulgaris*

Resident and winter visitor, breeds

In 1830, Dr E. Moore noted that the Starling was a common winter visitor which had been known to breed in Devon. D & M stated that it had greatly increased since about 1844 and was resident and breeding throughout the county except on Dartmoor, while in 1906 D'Urban described it as an extremely abundant resident, with vast winter flocks.

That it has continued to increase and spread is shown by its present status in the Postbridge area of Dartmoor, where Dare and Hamilton report that the breeding population increased from one or two pairs in 1956 to about thirty pairs by 1967, and a large but sporadic roost at Soussons Plantation has been formed since 1963.

From June until March, Starlings congregate in roosts from which foraging flocks disperse daily to feeding areas. Their numbers are vastly increased by winter visitors arriving mainly in October, and later by hard weather immigrants. Such roosts, which may be used for many successive years, or intermittently, include Sowton, with 300,000 birds in January 1932; Topsham reed beds 200,000 in December 1950; West Worlington 200,000 in January 1950; Farringdon 100,000 in February 1954; Combe Brake near Lympstone with over 1,000,000 in February 1955; Peamore near Exeter 500,000 in February 1964; and vast numbers annually at Halwill Forest.

Ringing recoveries in Devon include birds from Sweden, Lithuania, Germany, and Russia (36° E).

The Starling is abundant and regular on spring and autumn passage on Lundy, where flocks of about 10,000 were recorded in November 1953 and November 1959, and varying numbers are present throughout the year. Breeding is irregular: two pairs around 1909, one pair in 1938 and 1942-3, two pairs in 1962, one in 1964, and about fourteen pairs in 1966.

ROSE-COLOURED STARLING   *Sturnus roseus*

Vagrant

D & M refer to this species as 'a casual visitor, of rather frequent occurrence, principally to the South Hams and to Lundy Island, during the spring and summer months'. About sixteen records are listed of birds obtained in the county during the last century. The occurrences in Devon of this central Asiatic species have been only half that number during the present century and, except for one on Lundy, all have been in the southern part of the county.

In 1932 an adult male was seen in a garden at Exmouth by Colonel R. M. Byne (*BB* 26 : 170); and another was recorded at Stockland on 7-8 May 1937 by F. C. Butters, who also saw an adult male at Seaton on 28-30 July 1945, which he recorded in *BB* 38 : 373. In the same year two were reported at West Hill, Ottery St Mary on 16 May. During July 1947 an adult visited a bird table at Loddiswell and, what was quite likely the same bird, was seen at Kingsbridge on 11 August. A juvenile, which was detected by R. G. Adams on a refuse dump at Exmouth on 15 October 1950, is well documented in the *Devon Report*. Lastly, another juvenile occurred on 12 October 1962 at Kingswear where it was observed by M. R. Edmonds, whose detailed description of the bird is given in the *Devon Report* for 1962. This record was accepted by the Rarities Committee and communicated in BB 56:407, where it was noted that Rose-coloured Starlings are frequently imported as cage-birds.

Although this species was evidently a frequent visitor to Lundy during the third quarter of the last century (*BL* 98), it is known to have occurred there only once during the present, when an adult male was observed from 18 to 26 June 1934, and was reported in *BB* 28 : 49 by H. H. Davis, and *BB* 28 : 77 by David Lack, who saw it on 24 and 26 June.

MYRTLE WARBLER   *Dendroica coronata*

Vagrant

There are only two known occurrences of this American warbler in Britain; both were recorded in Devon. The first Myrtle Warbler occurred in 1955, when it regularly visited a bird table in the garden of Dr and Mrs Cook at Newton St Cyres. It was first seen on 5 January by David Cook, and remained in the vicinity until 10 February when it was found dead in their garden. During this period the bird was very closely studied by many ornithologists, through the kindness of Mrs Cook, and was photographed by E. H. Ware,

whose illustrations support the very full account in *BB* 47 : 204-7 by
F. R. Smith, who identified the bird. I was fortunate enough to watch
it myself on 12 January and on a subsequent occasion. The bird
appeared to be in perfect condition at the time of its death; its skin
was later mounted and is preserved in the Royal Albert Memorial
Museum at Exeter.

The second example, an immature bird, was identified on Lundy
by W. B. Workman on 5 November 1960. It was trapped and ringed
on 8 November, when a detailed description and colour photographs
were taken before its release. The bird remained on the island until
14 November, during which period, and whilst in the hand, it was
examined by other observers. W. B. Workman, whose account of
the occurrence appeared in *BB* 54 : 250-51, adds that there are at least
two records of Myrtle Warblers crossing the Atlantic on board ship
to within sight of the coast of Ireland or Britain.

YELLOWTHROAT   *Geothlypis trichas*
Vagrant

The trapping of a Yellowthroat on Lundy on 4 November 1954 constitutes the first and only record for Britain of this American warbler.
Barbara Whitaker in her fully documented account of this occurrence, in *BB* 48 : 145-7, describes it as superficially like a small leaf
warbler, though certain of its movements were reminiscent of a
Wren. The bird, a first winter male, was kept at the Observatory
overnight and, after a laboratory description had been taken, was
ringed and released, having been examined by several of the islanders
including F. W. Gade and John Ogilvie. From the details obtained it
was impossible to assign it to any particular one of the twelve subspecies recognised by the American Ornithologists' Union. A common species in North America, the Lundy bird is assumed to have
been driven across the Atlantic by adverse winds during its southwards migration.

BALTIMORE ORIOLE   *Icterus galbula*
Vagrant

The first accepted British record of this American species was an
immature female which was present on Lundy from 2 to 9 October

1958. It was trapped on 2 October by R. H. Dennis and W. B. Workman, whose account of the occurrence and laboratory description of the bird are given in *BB* 56 : 52-3.

Two further examples of this trans-Atlantic drift migrant occurred on Lundy on 17 October 1967, *BB* 61 : 356.

HAWFINCH   *Coccothraustes coccothraustes*

Resident and winter visitor, breeds

A rare and irregular breeding species, the Hawfinch is also a scarce winter visitor and passage migrant. It has been recorded in all parts of the county except on high moorland, and has occurred annually since 1930, but in general is more frequent in the south and east than elsewhere. D & M described it as an irregular winter visitor which had been known to nest at Loddiswell near Kingsbridge. D'Urban's MS adds that a nest and eggs were obtained at Cowley Bridge in 1906 and young were seen at Exmouth in June 1920.

Breeding records listed in the *Devon Reports* are Lympstone 1935, Budleigh Salterton 1936, Lympstone 1937, Newton Abbot 1940, Lympstone 1942-3, Exton 1943, Budleigh Salterton 1944, Lympstone (two pairs) 1948, Bicton 1949, Plymouth 1953-4, Lympstone 1954, Longdown and Shobrooke 1956, Southleigh 1959, and Newton Abbot 1962. In addition breeding behaviour has been observed in a number of other years, particularly at Lympstone, and at Plymouth during the 1950s.

Records of more than usual include ten at Lympstone in December 1940, eight in November 1944, and eleven in September 1949; eleven at Netherton in January 1950, thirteen at Budleigh Salterton in March 1951, and up to twenty-six at Plymouth during April 1953. In this locality in 1954, nineteen were seen on 30 January and twenty-three on 28 March, while in 1957 twelve occurred there on 22 March. Six were observed at Teignmouth in July 1963. There is often a noticeable spring passage in March and April, and evidence of movement in October. Although the Hawfinch has increased during the present century, particularly 1940-60, it has been scarce since 1964. Recently-fledged young seen on Lundy in 1927 constitute the only evidence of breeding there. Ones and twos have been recorded on the island in eight of the years 1947-66, mostly between June and November.

GREENFINCH   *Carduelis chloris*

Resident and winter visitor, breeds

In the breeding season the Greenfinch is very much a bird of gardens, shrubberies, and the edges of woods, but during the autumn and winter many resort to stubble fields, arable land, reed beds, saltings, and sandy shores. Widespread and plentiful throughout Devon, except on high ground, it suffered great losses in the winter of 1963, from which it has probably now recovered. In the Postbridge area of Dartmoor P. J. Dare regards it as scarce and occurring in small numbers, mainly in winter.

Large flocks that gather on the coast in winter include one of at least 700 on the Teign estuary in January 1948; 300 at Slapton in December 1949 and the same number at Starcross in January 1953; over 200 on the shore at Dawlish Warren in September 1960; and regular records of smaller flocks in many coastal and inland localities.

On Lundy, where the species is recorded as having nested in 1934 and 1938, small numbers normally occur on spring and autumn passages. A peak of twenty-five in mid-November 1952 appears to be the most recorded.

GOLDFINCH   *Carduelis carduelis*

Resident, breeds

The Goldfinch, according to D & M, was formerly numerous but had become rather scarce by the end of the nineteenth century, largely due to the great numbers taken as cage-birds, but also to the incidence of hard winters. With the cessation of bird-catching it has greatly increased during the present century and is well distributed in all areas, except on high moorland. As a breeding bird it occurs in gardens, orchards, and cultivated land, but in autumn and winter foraging flocks occur in many types of open and waste ground, wherever there are thistles. On Dartmoor it is thinly distributed in semi-cultivated ground, a few breeding pairs and autumn birds being reported in the Postbridge area.

Although reductions in population followed the severe winters of 1947 and 1963, the Goldfinch has been fairly quick to recover, and the usual autumn flocks have been reported within a few years.

Flocks of more than the 100 or so usually seen include 300-400 on Woodbury Common in October 1941; a single flock of almost 500 at Lympstone in September 1942; and about 400 each at Woodbury in September and Plymouth in December 1957. In February 1965 a flock of 200 was seen at Turf, and several flocks of 150 have been recorded from different localities since 1963.

Small numbers have been recorded on Lundy in all months, but it is more regular on spring and autumn passages, with up to twenty in October, and a maximum of thirty on 19 October 1966.

Breeding was reported in 1908, possibly 1909, one pair in 1922 and 1928, and one definitely and another probably in 1959.

SISKIN *Carduelis spinus*

Resident and winter visitor, breeds

The Siskin nowadays occurs more frequently and in larger numbers than were recorded by D & M, who described it as of uncertain appearance and occurring sometimes in severe winters. In addition, it has become established as a breeding bird on Dartmoor since nesting was first proved in 1957.

The records show that since 1930 it has occurred annually in winter in fluctuating numbers, and is quite widespread in some years though in general it tends to be more frequent in the south and east than in the north and west. Between October and March flocks of up to about thirty frequent the alders and birches along the river valleys, and it occurs fairly regularly along the banks of the Exe at Topsham, where the birds feed amongst the debris on the tideline.

Flocks of around fifty have been recorded in the Otter valley in November 1933, Lympstone in December 1941, Budleigh Salterton in March 1949, Dawlish Warren in January 1956, Topsham in January 1960, and Crediton in December 1963. About 100 were seen beside the River Otter in December 1966, and a flock of 100 at Stover on 14 April 1956 were evidently on spring passage.

The discovery of Siskins breeding in the conifer plantations at Bellever was made by P. J. Dare in 1957, when a pair was observed feeding young (*BB* 55 : 193-5). Successful breeding also occurred there in 1958 and 1961 and possibly other years. As Siskins have also been recorded fairly regularly during May/July at Fernworthy since 1957, and occasionally at other plantations, it is presumed that they now nest annually.

On Lundy a pair bred successfully in 1952. The species normally occurs there as an irregular winter visitor and passage migrant, being scarce or absent in some years, occurring more frequently in autumn than spring, and exceptionally arriving in large numbers, as in 1959 when peaks of 200 on 24 October and 250 on 25 October were recorded.

LINNET *Acanthis cannabina*

Resident and winter visitor, breeds

An abundant and widespread species, the vivacious Linnet is present throughout the year. During the breeding season it occurs chiefly on gorse-clad commons and rough ground, but also in a variety of habitats having suitable cover throughout the county, while in winter it deserts these areas for cultivated ground, stubble fields, the sea shore, and the tideline of estuaries. It is a common breeding bird from the coast to the moorland, wherever suitable habitats exist— whether a small patch of roadside gorse or a large tract of open common. On Dartmoor it breeds chiefly in the areas where gorse abounds, but in winter is more widespread.

Described by D & M as 'resident, generally distributed and abundant', its status has evidently not changed to any noticeable extent, except that its numbers were reduced, particularly on Dartmoor, by the severe winter of 1962-3, a loss which has since been made up.

On Lundy, where it nests regularly, the breeding population fluctuates, being described by Loyd as very numerous in 1922-3, by Wynne-Edwards and Harrison as thirty-eight pairs in 1930, and by Perry as six pairs in 1939. The *Lundy Reports* give the number as about twenty for the years 1952 and 1953 and twenty to twenty-five in 1962, but following that severe winter it was reduced to only a few pairs in 1963 and 1964. Peter Davis states that the breeding birds arrive on the island during late March and April, and leave in September and October. Passage migration is recorded on Lundy in both spring and autumn, but greater numbers of birds are involved during the latter season.

On the mainland, large flocks, several hundred strong, occur during the autumn and winter in all areas, but particularly in coastal and estuarine localities. A flock of about 500 observed at Start Point on 3 October is typical of the autumn movement.

TWITE   *Acanthis flavirostris*

Vagrant, has bred

D & M assessed the status of the Twite in Devon as 'a casual visitor of very rare occurrence in south Devon, where the only known instances are not well authenticated but not unfrequently seen in the northern part of the county'. They also stated that flocks used to visit north Devon in autumn and winter and that they once killed six or eight Twites out of a flock near Barnstaple. *The Handbook* says of this species, 'small colony reported nesting N. Devon 1904, but apparently not now'. I also have a letter dated 10 September 1937 from H. F. Witherby in which he assures me that 'the breeding of this species was certainly correct'. D'Urban's MS records that a nest and eggs of the Twite, taken in north Devon, were exhibited at a meeting of the British Ornithologists' Club on 15 June 1904.

The status of this species in the county is also treated in some detail in vol 1, p 143, of D. A. Bannerman's outstanding work *The Birds of the British Isles*, in which he says that L. R. W. Loyd assured him that several pairs of Twites were resident on high ground between Sidmouth and Exeter some time before 1929. In his *BSED*, Loyd writes 'we have ourselves found it regularly in one locality in the south-east where there is a strong possibility of its nesting in small numbers, and he mentions, amongst other records, that W. Walmesley White observed eleven or twelve Twites near Budleigh Salterton on 5 January 1918.

Whatever the exact status of this species may have been in the early 1900s, it is most certainly a rare bird in Devon at the present time, its disappearance being consistent with the general decrease and retreat northwards to which J. L. F. Parsloe refers in *BB* 61 : 54. Two of the only three records of this species mentioned in the *Devon Reports* are so lacking in detail that they cannot at this stage be regarded as really authentic. They comprise a single bird at Little Haldon on 1 May 1934, a party of six at Woolacombe on 26 October 1937, and the third and most recent record for the mainland (*Devon*

Report for 1950), a flock of about eight birds on Bursdon Moor on 10 October 1950, two of which were very closely watched by R. L. Winter within a few feet of his car.

Omitting previous generalisations about its status on Lundy, the records indicate that the Twite has been recorded there on five occasions since 1937. A party of six was observed during the spring of that year by F. W. Gade (*BL* 102); and the *Lundy Reports* record two birds on 13 October 1952, one on 12 and 13 February 1953, a male on 5 June 1954, and a female on 6 May 1958.

Since writing this account, I am informed by F. R. Smith that he closely watched a party of five on Woodbury Common on 14 December 1968.

LESSER REDPOLL   *Acanthis flammea*

Resident and winter visitor, breeds

Until the 1950s the Lesser Redpoll was an almost annual but uncommon winter visitor, occurring in small parties principally along the river valleys of east Devon and occasionally further west; an unusual record was a flock of forty-one at Lapford in December 1931. The occurrences became rather more frequent during the late 1940s, with flocks of up to thirty being observed at many places.

In 1952-3 ones and twos were recorded during June at Bovey Tracey and Fernworthy, and a pair with young were seen at Brendon Forest in August 1953. Breeding was proved at Woodbury Common in 1954 and near Bovey Tracey 1954-6. During the breeding seasons 1957-62 pairs and singing males were reported in practically every State forest as far west as Hartland, and the species was quickly becoming established in all parts of the county, but the hard winter of 1962-3 severely reduced its numbers.

Although breeding has since occurred annually in many localities, usually conifer plantations, the numbers are not yet up to the 1962 level.

Recent winter and spring flocks include fifty at Merton in April 1965 and fifty-seven at Danes Wood in March 1966.

Sporadic breeding at Lynton in 1879 is mentioned by Pidsley, and was proved by B. G. Lampard-Vachell at Weare Giffard in June 1936.

Very small numbers have been recorded on Lundy in spring or

S

autumn in twelve of the years 1947-66.

The Mealy Redpoll (*A.f. flammea*) is listed by D & M, who state that one obtained in north Devon came into their possession. The *Devon Reports* mention occurrences in December 1931 and April 1944.

SERIN   *Serinus serinus*

Vagrant

The only reference by D & M to this species is on p 436 where a record is included in their *addenda*. It relates to a single bird which was obtained by a bird-catcher between Exmouth and Budleigh Salterton on 29 November 1891, and although D'Urban thought that it was possibly an escaped bird, the record was evidently considered by Witherby to be satisfactory, as it was included in *The Handbook*.

There are five further records of this small vivacious finch during the present century, most of them in quite recent years. A party of four is reported to have been seen at close quarters by F. W. Gade on Lundy on 21 April 1943 (*BL*:102). A female was present on the island from 28 July to 24 October 1956, and a first-winter male was seen there on 23 April 1959, as stated in the *Lundy Reports* for 1956 and 1959-60. Since then, there have been two reliable sight-records for the mainland.

In the *Devon Report* for 1964 R. Khan gives a detailed description of one that frequented a garden at Budleigh Salterton from 22 to 25 May of that year. Although the record was accepted by the Rarities Committee and noted in *BB* 58:369, they considered that the bird might have escaped from captivity. Lastly, two Serins were seen at Slapton on 18 and 20 December 1966 by L. I. Hamilton and F. R. Smith and one on 24 December by M. R. Edmonds and R. M. Curber, both sightings being detailed in the *Devon Report* for 1966. This record was accepted by the Rarities Committee and published in *BB* 60:330, where it is remarked that an unprecedented number of Serins was reported in Britain during 1966, following the gradual northward extension of its range in Europe during the past 150 years. This was followed by the announcement in *BB* 61:56 that the species bred in southern England in 1967, and that further colonisation of Britain seemed likely.

A further record of one seen by P. F. Goodfellow at Roborough, Plymouth on 1-2 August 1968 has been accepted by the Rarities Committee.

BULLFINCH *Pyrrhula pyrrhula*
Resident, breeds

A common and widespread breeding bird, the Bullfinch is resident throughout the year and, according to the frequent statements in the *Devon Reports*, has increased everywhere. From its numbers in all the conifer plantations, there is no doubt that it has benefitted by afforestation. There are breeding records from localities throughout the county, including western Exmoor and Dartmoor; on the latter it nests in the plantations at Bellever and Fernworthy and elsewhere, and it also nests in the Postbridge district and the West Webburn valley, though in smaller numbers than at lower altitudes.

During winter, small parties of up to ten are frequently encountered in all areas, but there are also records of flocks of up to twenty from widely scattered localities, and one exceptionally large flock of about 126 was recorded by G. H. Gush at Eastleigh near Bideford on 14 February 1936. An influx in the autumn has been noted in several recent years at Slapton, and that some movement does occur is shown by the occasional records of this species on Lundy. It has been recorded there on approximately twenty occasions since 1947, when mostly singles but up to three have occurred during the months of March and April and September to November.

SCARLET ROSEFINCH *Carpodacus erythrinus*
Vagrant

The Scarlet Rosefinch, or Grosbeak, a bird of eastern Europe, has occurred three times in the county, three examples having been trapped on Lundy in recent years. The first occurrence, as stated in the *Lundy Report* for 1959-60, was a female or first-winter male which was mist-netted on the island on 10 September 1959, and was present until 13 September. The second example was trapped by C. S. Waller on 7 September 1966 and the record was accepted by the Rarities Committee and listed in *BB* 60:331. The third was trapped on 22 October 1967, *BB* 61:357.

PINE GROSBEAK *Pinicola enucleator*
Vagrant

The inclusion of this vagrant from the forests of Scandinavia and northern Russia rests on a single sight-record of a female seen on Lundy on 7 May 1958 by B. K. Whitaker, at that time Warden of the Observatory. The account of this occurrence, contained in the *Lundy Report* for 1958, gives only the bare facts and omits even a brief description of the bird and the circumstances under which it was observed.

The record of a bird of this species, said to have been killed near Exeter in the winter of 1854-5, was rejected by D & M, who knew of no valid record of this large finch.

CROSSBILL *Loxia curvirostra*
Irregular visitor, breeds sporadically

Following the periodic irruption of Crossbills from northern Europe in late summer, the birds occasionally remain for a year or so and breed, but they gradually die out. D & M recorded breeding near Newton Abbot in 1839 and at Hatherleigh in 1894.

During the present century irruptions occurred in 1927, 1935, 1942, 1953, 1956, 1958, 1959, 1962, 1963, and 1966. The species has been recorded in Devon in all but seven of the years 1928-67, the exceptions being 1932-4, 1948, 1951, 1952, and 1955. Breeding by a few pairs was recorded in the Woodbury Common area in 1936-7, 1939-43, 1957, and 1960-61. It was also reported at Lapford in 1923, and probably occurred at other localities in recent years. The irruptions usually occur in the months of June to August, and in some years Crossbills are recorded throughout the county, almost always in coniferous woods, in flocks of up to about forty, while in years of small invasions they may only reach south-east Devon.

Some of the larger numbers recorded since 1928 are Rockbeare Hill, sixty in July 1935; Woodbury Common, flocks of up to sixty in June 1953; Haldon, seventy-five in July 1953; Woodbury Common, 100, and Torquay, fifty in July 1958; Woodbury Common, fifty and sixty in July 1962; and Clifford Bridge, 100 in August 1963. And in July 1966 East Budleigh Common had ninety, Ashclyst Forest had over fifty, Haldon had fifty, and Clifford Bridge over fifty.

Flocks of up to fifty have been recorded in most of the Dartmoor conifer plantations and at Halwill, Hartland, Holsworthy, and Plym Forests.

Apart from earlier records, Crossbills have been observed on Lundy in nine of the years 1947-66, all between June and October, with a maximum of thirty-five in July/August 1959.

PARROT CROSSBILL *Loxia pytyopsittacus*
Vagrant

The Handbook admits one Devonshire record of the Parrot Crossbill, a male of which was obtained at Marley near Exmouth in 1892. This specimen, which was critically examined by Willoughby Lowe, is preserved in the Exeter Museum. Although D & M do not mention this date, they state that 'in January 1888 a large flock visited Marley, near Exmouth, remaining for several weeks. Some specimens were shot, which we examined in the flesh, and are undoubted examples of this larger race'. Whilst it is impossible now to verify the identification of these birds, the species has been included on the strength of the 1892 record which Witherby accepted.

TWO-BARRED CROSSBILL *Loxia leucoptera*
Vagrant

D & M state that a male American White-winged Crossbill was picked up dead on the shore at Exmouth on 17 September 1845 by E. B. Fritton, who reported the facts in *The Zoologist*, 1845, p 1190. A strong south-west wind had been blowing for several days.

This record, which refers to the American race (*L.l. leucoptera*) of the Two-barred Crossbill (*L.l. bifasciata*), was included in W. B. Alexander and R. S. R. Fitter's paper 'American Land Birds in Western Europe', *BB* 48 : 1-14. No occurrences of the American subspecies were accepted by the editors of *The Handbook*, who doubted whether the birds occurred in a truly wild state.

RUFOUS-SIDED TOWHEE *Pipilo erythrophthalmus*

Vagrant

In recording the occurrence of this species on Lundy on 7 June 1966 the Rarities Committee stated, in *BB* 60 : 332, 'This is the first British record of this North American finch, which breeds from southern Canada to Central America and winters in the southern part of its range'. The bird, which was trapped, was examined in the hand by Miss J. Mundy, J. Ogilvie, and C. S. Waller.

CHAFFINCH *Fringilla coelebs*

Resident, winter visitor and passage migrant; breeds

An extremely widespread and abundant resident, the Chaffinch breeds throughout the county except on high open moorland. On Dartmoor it nests commonly up to the tree limit, at about 1,400 ft. In an area of approximately 15 square miles around Postbridge, the breeding population during the past decade was estimated by Dare and Hamilton as 250-500 pairs, with no apparent reduction by the severe winter of 1962-3.

As a winter visitor it occurs abundantly in flocks of up to about 400 in all areas, but particularly in stubble and root fields. An unusually large gathering of 1,500 was recorded on Dartmoor in February 1957.

Considerable migratory movements are observed in some years on the north coast, chiefly in the Bull Point and Hartland Point areas, during late October, often involving some thousands of birds, which Dr D. Lack considers to be heading for Ireland (*BB* 50 : 10-19).

From three to twelve pairs breed annually on Lundy. A light spring passage occurs there from February to April, but heavy autumn movements from September to November, involving thousands of birds, are recorded in some years, as in October/November 1951, and 26 October 1963, when a peak of 6,000 moved south. Davis (*BL* 104) states that both the north and central European forms

F.c. coelebs and *F.c. hortensis* occur.

There does not appear to have been any significant change of status since last century.

BRAMBLING *Fringilla montifringilla*

Winter visitor

Although an annual winter visitor, the Brambling fluctuates widely in numbers, being abundant in some years but scarce in others. Visiting stubble, kale and root fields, rickyards and beech woods, it occurs in all parts of the county, normally in small numbers of up to about fifty, in mixed flocks of finches, but not infrequently in flocks of 100-200 and occasionally more. Although the first birds are reported in early October, it is usually from December to February that the larger flocks are seen, and sometimes early April before the last birds leave.

During the past 30 years flocks of 100 have been recorded from widely separated localities throughout Devon, while larger numbers of around 200 have been reported from Black Torrington in November 1940, Lympstone in January 1941, Postbridge in February 1957, Crediton in January 1965, Bickington (near Ashburton) in December 1966, and Boreston, Crediton, and Cheriton Fitzpaine in January/ February 1967. The largest number recorded was a flock of some 500 at Cockwood in late December 1957 and subsequent weeks.

Migratory movements of Bramblings, greatly outnumbered by Chaffinches, have been observed at Bull Point on the north coast. Dr D. Lack reported that hundreds passed west-north-west in late October 1956 (*BB* 50:10-19) and A. J. Vickery reported movements of several hundred on 25 and 27 October 1963.

The species has been recorded almost annually on Lundy since 1949, mostly on spring and autumn passages, but sometimes wintering, and in all months from October to April, with peak numbers of usually up to about forty but with 206 on 25 October 1955.

YELLOWHAMMER *Emberiza citrinella*

Resident, breeds

An abundant and beautiful bird, the Yellowhammer has a wide breeding distribution throughout the county, including high moorland with sufficient bushes; but it is absent from completely open moorland and from some low-lying ground where in places it is, or was until 1963, replaced by the Cirl Bunting. Unlike the latter, it survived the early months of 1963 far better than many other species, and was still quite plentiful at the end of the year, when a flock of ninety-two was recorded at Start Point on 22 September and another of 100 on Beaford Moor in early March 1964. With about thirty breeding pairs on a farm at East Boreston in 1965, it was stated to be the commonest breeding species.

Winter flocks of up to about fifty birds are regularly recorded on agricultural land in all areas.

Loyd stated that about eight pairs nested on Lundy in 1922; two pairs almost certainly bred in 1951, and a pair was present throughout the summer of 1952, but there was no satisfactory evidence of breeding then or in any subsequent year. Single birds or occasional parties of up to five or six are recorded on the island every year, usually in spring and autumn, but irregularly in all months.

There is no evidence of any significant change of status in Devon since the last century.

CORN BUNTING *Emberiza calandra*

Resident, breeds irregularly

D & M described the Corn Bunting as being resident but local and confined principally to the coasts, but nowhere numerous. Loyd considered it rare in east Devon, but W. Walmesley White recorded that two pairs bred regularly at Budleigh Salterton during the 1920s.

Until about 1940 a few pairs bred annually at Braunton, where I regularly used to see them, and I recorded singing males at Bigbury in 1936. The species bred regularly at Roundeswell, Barnstaple until about 1945, since when a few single birds were reported up to 1954. One pair bred successfully at Saunton in 1957, but the only mainland records since this are of singles at Lympstone in March 1958, Saunton in August 1959, and Challaborough in August 1966.

Davis gave about seven records for Lundy between 1933 and 1951, since when one was seen in May 1954 and another in April 1965. Although the Corn Bunting still breeds regularly in the west of England, it has decreased somewhat in Cornwall and more so in

Somerset, if not in Dorset.

A pair almost certainly bred at Challaborough in 1968, and breeding probably occurred near Hartland Point during the early 1960s.

BLACK-HEADED BUNTING   *Emberiza melanocephala*

Vagrant

The inclusion of this rare straggler from south-eastern Europe, of which Bannerman in 1953 and Fisher in 1967 gave the total British occurrences as about fifteen, rests on a well authenticated record of a male detected by T. J. Richards on Salcombe Hill, Sidmouth on 4 and 6 October 1951. A full account of this sight-record is contained in *BB* 45:406, as well as in the *Devon Report* for 1951.

Although Loyd included this species in his *Lundy, its History and Natural History*, on the strength of Chanter's list, the record has been rejected as unacceptable by later writers. There is, however, a recent record for the island of a female seen on 20 and 22 September 1957. In the account of this occurrence in the *Lundy Report* for 1957, precise details are given in support of the identification, based on the examination of skins of females of this species in the British Museum. This record which relates to the year before the first report of the Rarities Committee does not appear to have found its way into *BB* journal.

CIRL BUNTING   *Emberiza cirlus*

Resident, breeds

Until the extremely severe winter of 1962-3, the Cirl Bunting was quite abundant in parts of the county, but during the first three months of 1963 was almost exterminated by the intense cold. Although a few pairs survived and the species has since made some progress, it has been unable to recover to anything like its former status, even after five years. However, previous fluctuations, though not on the same scale, were recorded by D & M, and it is to be hoped that this beautiful bunting will regain its lost ground.

A Mediterranean type, which reaches the northerly limit of its range in England, the Cirl Bunting may have colonised England only

T

at the end of the eighteenth century, according to Alexander and Lack (*BB* 38 : 45), as Gilbert White did not observe it at Selborne, and it was first met with in this country by Colonel Montagu, who discovered it at Kingsbridge in 1800. D & M described it as 'rather local but common in some places on the south coast, and near Barnstaple in the north'.

A survey of its distribution which was carried out by the Devon Bird-Watching Society in the three years from 1957 to 1959 showed that the Cirl Bunting was confined to a fairly narrow coastal strip extending along the entire south coast, the more low-lying parts of the north coast from Combe Martin westwards to Abbotsham, and the country bordering the lower reaches of the main rivers. The stronghold of the species is the lower Exe valley from around Bickleigh to the coast, with the Exe's tributaries, the Clyst, Culm, and Creedy—a country of pasture land with hedgerows and many elm trees. Similarly, the species extends some way up the Otter and the Teign, and as far as Buckfastleigh on the River Dart, and along the river valleys of the south-west, although there was no evidence of its occurring any distance up the Tamar valley. In the north it inhabits the low-lying country around Braunton and the lower reaches of the Taw and Torridge, but was not recorded by B. G. Lampard-Vachell in his list of the birds of Torrington. The species is absent from the whole of Dartmoor, but has been noted by H. G. Hurrell on the fringe at Wrangaton, and, except for a few small, isolated pockets, is not found in the remainder of inland Devon.

At the time of the enquiry, the results of which are contained in the *Devon Reports* from 1957 to 1959, I found the species particularly plentiful in the Clyst valley, where it appeared to replace the Yellowhammer in the low lying areas. At Dartington Hall, David Cabot reported seven breeding pairs in 1954 and five in 1956. Two older records, mentioned in *BB* 3 : 125 and 195, are also of interest : the first by Norman Gilroy who found the species 'in great abundance' in south Devon during August 1909, and the second by S. G. Cummings who confirmed this in south-east Devon during the same month and year.

The *Devon Report* for 1966 includes records from twenty-one localities all within the known range, and suggests that the recovery, although very gradual, is still continuing. Pre-1963 records of wintering flocks include twenty-three at the edge of the River Exe at Topsham during October 1949, up to thirty beside the Otter estuary during the months of October to December 1951, and about thirty at Christow in December 1953.

According to Peter Davis, the Cirl Bunting has been observed on Lundy on twelve occasions since 1900, including three September and one April record, all of singles, listed in the *Lundy Reports* for the four years from 1948 to 1951, but there is no subsequent record from there of this essentially sedentary bird.

ORTOLAN BUNTING *Emberiza hortulana*

Passage migrant

D & M knew of no occurrence of the Ortolan Bunting, which, even in *The Handbook*, is stated to be rarely recorded on the west side of England. Since the advent of the bird observatories, however, it has been found to be a scarce but regular autumn passage migrant in places where it was formerly quite unknown. The occurrence of one on Lundy on 11 May 1949 was the first record for Devon, since when it has been encountered every year except 1961 between the end of August and early October, but mainly during September. The numbers range from one in some years to parties of up to six in others, and a maximum of about nine records during the autumn of 1951. In some years small parties remain on the island for a few days or even weeks, as in 1965, when up to six were present from 13 to 27 September.

The records for the mainland, however, are extremely few, and comprise only two individual birds, both in the south of the county —a male at Modbury on 30 April 1952 and another male, seen with a flock of Yellowhammers, at Prawle Point by J. R. Brock on 5 September 1965.

LITTLE BUNTING *Emberiza pusilla*

Vagrant

There was no Devon record of this north-east European and Siberian species until 1937, when two were observed at an unspecified locality in east Devon on 11 November. The description of the birds by two observers is included in the *Devon Report* for 1937. The only other record of this very small bunting relates to two adult males seen together on Lundy, one of which was trapped and ringed on 16 October 1951 and seen again on 19 October. Peter Davis's detailed account of this occurrence is given on p 19 of the *Lundy Report* for 1951. The Little Bunting, although very rare in the south-west of England, occurs regularly on Fair Isle, mainly on autumn passage, and less frequently at other points, chiefly on the east side of Britain.

REED BUNTING *Emberiza schoeniclus*

Resident, breeds

A fairly common resident, the Reed Bunting breeds on the estuarine marshes and reed beds, along river valleys, beside lakes, on upland marshes and moorland bogs, and also in young conifer plantations. On Dartmoor it is well distributed in marshy areas during the breeding season, and has been recorded in June at an altitude of 1,775 ft near Dart Head; and breeding has been confirmed at 1,500 ft on Exmoor. Nesting in young conifer plantations, often some distance from water, has been recorded at Fernworthy and Soussons, and some of the more favoured breeding localities are Slapton Ley, Braunton and Chivenor Marshes, and the Axe and Exe estuaries. Breeding has also been recorded at Burrator, Tamar Lake, Wistland-pound, and many other localities throughout Devon.

Winter flocks of twenty or thirty birds are regularly recorded in many coastal and inland localities, the former suggestive of some migratory movement, eg a flock of twenty at Start Point in January 1958, and fifty at Slapton in March 1966. A flock of more than usual, on East Budleigh Common in February 1967, comprised sixty birds.

Singles and small parties of up to four have been recorded on Lundy in 11 of the 20 years from 1947, mostly during the months of February to May and October/November.

There is nothing in the accounts of Pidsley, D & M, and Loyd to suggest any marked change in the status of this species, though it may possibly have increased slightly.

LAPLAND BUNTING *Calcarius lapponicus*

Passage migrant

Systematic watching on Lundy in post-war years has revealed that the Lapland Bunting, a species not reliably recorded in Devon until

1942, is a regular autumn passage migrant through the island. It has also been recorded as occasionally wintering both on Lundy and the mainland. No occurrence was known to any of the earlier writers, and the first mention of the species is a second-hand record of one supposed to have been seen on the Chagford side of Dartmoor in January 1934, but which is not properly documented.

The first authentic record is of a single bird identified on Lundy by W. B. Alexander on 10 September 1942 and reported in *BB* 36:140 as the first record for Devon. The same ornithologist, visiting the island again in 1948, detected another example on 8 September, which I also saw, and two or three others on subsequent days up to 14 September. Since 1948 this northern bunting has been observed on the island every autumn between the beginning of September and the end of November, but principally during September. The numbers are usually small and comprise a few single birds, but a party of six was present from 3 to 15 September 1949, with seven on 7 September, while maxima of five were recorded on 20 September 1956 and 8 September 1957.

The year 1953 was an exceptional one for the influx of Lapland Buntings into Britain. As far as Lundy is concerned, the first appeared on 3 September after which some were seen almost daily until 18 October, the peak numbers being at least seventeen on 5 September, twenty-three or more on 10 September, thirty-three on 15 September and eleven on 27 September. Single birds were seen up to 17 November. In a most interesting account of this invasion, in the *Lundy Report* for 1953, Peter Davis considers that these birds, over a hundred of which were thought to have passed through Lundy, were drift migrants from Greenland. The only winter record for the island is a single bird which was present during very cold weather from 2 to 16 February 1954.

There are several records for the mainland, all in recent years, namely a pair on Northam Burrows on 2 October 1961, one at Crow Point on the opposite side of the Taw estuary on 12 November, up to seven birds on the Braunton Pill between 17 December 1961 and 10 January 1962, two seen on Dawlish Warren on 20 December 1964, and one at Tavistock on 8 November 1967—all of them listed in the *Devon Reports*.

SNOW BUNTING   *Plectrophenax nivalis*

Passage migrant and winter visitor

A passage migrant and winter visitor, the Snow Bunting occurs mainly on the sandy coasts at Northam and Braunton Burrows, Dawlish Warren, and Slapton, but there are irregular inland records for Dartmoor and Exmoor and other high ground and also coastal areas.

Small numbers have been recorded on the mainland in all but four years since 1931. The approximately ninety records involve about 230 birds, of which fifty-four occurred in October, seventy in November, fifteen in December, forty-eight in January, nineteen each in February and March, and a few in other months. Almost eighty records refer to singles and occasional parties of up to four. The main records are a flock of seventeen on Braunton Burrows on 27 January 1945, thirteen on Northam Burrows on 5 November 1955, about fifteen at Slapton on 22 November 1947, eleven inland at Winkleigh on 26 February 1965, several flocks of up to nine on Northam Burrows during the 1960s, and seven or eight on Holdstone Down in March 1967. Singles occur with some regularity on Dawlish Warren.

Systematic observation on Lundy since 1947 has shown it to occur in small parties as a regular autumn passage migrant, less regularly in spring, and occasionally in winter. Of the minimum of 250 birds recorded 1947-66, all occurred between September and May, with three-quarters of the total in October and November. Peaks of thirty-two occurred on 11 October and forty-two on 9 November 1950, and twenty-four during November 1957.

In addition to the mainland records stated above, no less than fifty-two Snow Buntings were seen on Northam Burrows by W. H. Tucker on 3 December 1968.

HOUSE SPARROW   *Passer domesticus*

Resident, breeds

An extremely abundant species, the House Sparrow is resident and breeds wherever there are human habitations, except on some of the more remote farms on Dartmoor, and possibly west Exmoor. It is recorded, however, in such moorland villages as Postbridge and Widecombe and occurs as a summer resident, in small numbers, on some of the less isolated farms. At his home on the southern edge of Dartmoor, H. G. Hurrell recorded a pair for a few months from February 1957, the first in 20 years, but they did not nest.

On Lundy there were no breeding birds in D & M's time, but D'Urban recorded in the *VCH* that a colony had recently become established. Loyd noted five or six pairs in 1922 and Wynne-Edwards recorded twenty-two pairs in 1930, while in 1939 Perry reported forty pairs. After an attempt to exterminate them, there was only one pair in 1942, and none has bred since. The *Lundy Reports* contain some twenty records for 14 of the 20 years from 1947 to 1966. Most occurred during April, May, and October, with a maximum of four birds on 18 October 1958.

Although there is little evidence of migration, apart from the Lundy records, a few were noted putting out to sea from Beer Head in September 1961 but repeatedly turning back.

SPANISH SPARROW   *Passer hispaniolensis*

Vagrant

An example of this Mediterranean and south-west Asian species which was recorded on Lundy on 9 June 1966 constitutes the first and only British record. The bird was observed by F. W. Gade, J. Ogilvie, and C. S. Waller, and the record was accepted by the Rarities Committee, in *BB* 60 : 333.

TREE SPARROW   *Passer montanus*

Winter visitor, breeds sporadically

Recorded in almost every year since 1928 and probably overlooked in others, the Tree Sparrow occurs as a regular but uncommon winter visitor, consorting with flocks of finches in stubble fields and winter crops. It was regarded by D & M as a casual visitor, and was reported by Loyd as breeding at Sid Vale, Branscombe, and near Honiton about the 1920s. The *Devon Reports* contain breeding records for Burlescombe in 1933, Torrington in 1944, and Brendon in

1944 and 1954; and a small breeding colony was found near Start Point in 1968.

Small parties of up to four or five are frequently reported from September to March, but a flock of seventy occurred at Slapton in December 1933, and up to forty in the following winter. Other records of more than usual include twenty at Exmouth in November 1935, forty on Exminster Marshes in March 1962, up to twenty at Slapton in December 1964 and 1967, and fourteen at Prawle Point on 30 October 1967. It is almost certain, however, that the species occurs more frequently than the records suggest.

On Lundy, where it has been recorded in ten of the years 1947-66 it occurs as an irregular visitor, mainly in April to June, with twelve in June 1958 and up to sixteen in the autumn of 1962. A few pairs are reported to have bred in 1928 and 1932, one pair bred in 1961, and two each in 1962 and 1963.

Selected Bibliography

ALEXANDER, W. B. et al. 'Observations on the Breeding Birds of Lundy in 1942', *British Birds* 38:182-91. 1945

ALEXANDER, W. B. & FITTER, R. S. R. 'American Land Birds in Western Europe', *British Birds* 48:1-14. 1955.

ATKINSON-WILLES, G. L. (Ed). *Wildfowl in Great Britain*. London, 1963

BRITISH BIRDS. Vols 1-61. 1908-68

BRITISH ORNITHOLOGISTS' UNION. *Check-List of the Birds of Great Britain and Ireland*. London, 1952

BRITISH TRUST FOR ORNITHOLOGY. *Bird Study* 1954-67, Vols 1-14

BOYD, H. 'The "Wreck" of Leach's Petrels in the Autumn of 1952', *British Birds* 47:137-63. 1954

CAMPBELL, B. 'The British Breeding Distribution of the Pied Flycatcher, 1953-62', *Bird Study* 12:305-18. 1965

COTTRILL, R. T. & HILL, P. A. 'The Birds of the Axe Estuary', *Devon Birds* 19:38-53. 1966

DARE, P. J. 'Birds of the Postbridge Area', *Devon Birds* 11:22-32. 1958

DARE, P. J. 'Birds of the Postbridge Area—Additional Notes', *Devon Birds* 12:9-12. 1959

DARE, P. J. 'The Breeding and Wintering Populations of the Oystercatcher in the British Isles', Min. of Agric. Fisheries & Food. *Fishery Investigations* Series 2, Vol 25, No 5. 1966

DARE, P. J. & HAMILTON, L. I. 'Birds of the Postbridge Area, Dartmoor 1956-67', *Devon Birds* 21:22-31 & 64-79. 1968

DAVIS, P. *A List of the Birds of Lundy*. Exeter, 1954

DEVON BIRD-WATCHING & PRESERVATION SOC. *Devon Birds*, Vols 1-21. 1948-68

DEVON BIRD-WATCHING & PRESERVATION SOC. *Annual Reports*, 1928-67

DIXON, C. *Bird-Life in a Southern County*. London, 1899

D'URBAN, W. S. M. & MATHEW, M. A. *The Birds of Devon* (Second Ed). London, 1895

D'URBAN, W. S. M. Bird Section in *Victoria County History of the County of Devon*. London, 1906

HARRISON, T. H. 'The Birds of Lundy Island from 1922 to 1931', *British Birds* 25:212-19. 1932

HARVEY, L. A. & ST. LEGER-GORDON, D. *Dartmoor*. London, 1953

HENDY, E. W. *More About Birds*. London, 1950

HOLMES, D. P. 'Report on the Status of Wildfowl in Devon', *Devon Birds* 8:1-18. 1955

HOLMES, D. P. 'Second Report on the Status of Wildfowl in Devon', *Devon Birds*, Supplement 13:1-12. 1959

HURRELL, H. G. 'A Raven Roost in Devon', *British Birds* 49:28. 1956

HURRELL, L. 'Golden Plover and Dunlin on Dartmoor', *Devon Birds* 4:3-4. 1951

LAMPARD-VACHELL, B. G. *Wild Birds of Torrington and District*. 1943

LEWIS, S. *The Breeding Birds of Somerset and their Eggs*. Ilfracombe, c 1952

LOWE, W. P. 'The Bird Collections in the Royal Albert Memorial Museum, Exeter', *The Ibis* 65-75. 1939

LOYD, L. R. W. *Lundy, Its History and Natural History*. London, 1925

LOYD, L. R. W. *The Birds of South-East Devon*. London, 1929
LUNDY FIELD SOCIETY. *Annual Reports* 1947-66
MADGE, S. G. 'Birds of the Crediton District', *Devon Birds* 13:14-22. 1960
MONTAGU, G. *Ornithological Dictionary*. London, 1802
MONTAGU, G. *Supplement to the Ornithological Dictionary*. London, 1813
MOORE, E. 'On the Ornithology of the South of Devon', *Trans. of the Plymouth Institution*. 1830
MULLENS, W. H. et al. *A Geographical Bibliography of British Ornithology*. London, 1920
NEWTON, R. 'Braunton Bird Notes', *Devon Birds* 12:22-30. 1959
PALMER, E. M. & BALLANCE, D. K. *The Birds of Somerset*. London, 1968
PALMER, M. G. (Ed). *Ilfracombe Fauna and Flora* (Bird Section by N. V. Allen). Exeter, 1946
PARSLOE, J. L. F. 'Changes in Status Among Breeding Birds in Britain and Ireland', *British Birds*. Vols 60 & 61. 1967-8
PERRY, R. *Lundy, Isle of Puffins*. London, 1940
PETERSON, R., MOUNTFORT, G. & HOLLOM, P. A. D. *A Field Guide to the Birds of Britain and Europe*. (Revised Ed). London, 1966
PIDSLEY, W. E. H. *The Birds of Devonshire*. London, 1891
RYVES, B. H. *Bird Life in Cornwall*. London, 1948
SLAPTON BIRD OBSERVATORY. *Annual Reports* 1960-67
SMITH, F. R. 'Myrtle Warbler in Devon: a new British Bird', *British Birds* 48:204, 1955
SMITH, F. R. 'Montagu's Harrier in Devon', *Devon Birds* 9:43-5. 1956
TAYLOR, I. & VICKERY, A. J. 'Birds of Wistlandpound Reservoir', *Devon Birds* 14:3-8. 1961
THOMPSON, W. H. *Devon, A Survey of its Coast, Moors & Rivers*. London, 1932
WHITE, W. Walmesley. *Bird Life in Devon*. London, 1931
WITHERBY, H. F. et al. *The Handbook of British Birds*. London, 1938-41
WRIGHT, F. R. Elliston. *Braunton, A Few Nature Notes* (Revised Ed). Barnstaple, 1932
WYNNE-EDWARDS, V. C. & HARRISON, T. H. 'A Bird Census on Lundy Island (1930)', *The Journal of Ecology* 20:371-9. 1932

Index

Index of English names mentioned in the Systematic List. Subspecies identifiable in the field are included, but for other subspecies see under the relevant species.